PHYSICS RESEARCH AND TECHNOLOGY

GLOBAL GEOMETRY OF SPACE-TIMES WITH SHELLS

PHYSICS RESEARCH AND TECHNOLOGY

Additional books in this series can be found on Nova's website
under the Series tab.

Additional e-books in this series can be found on Nova's website
under the e-book tab.

PHYSICS RESEARCH AND TECHNOLOGY

GLOBAL GEOMETRY OF SPACE-TIMES WITH SHELLS

VICTOR BEREZIN
AND
VYACHESLAV DOKUCHAEV

nova publishers
New York

Library of Congress Cataloging-in-Publication Data

ISBN: 978-1-63321-770-6

Published by Nova Science Publishers, Inc. † New York

Acknowledgments

Authors acknowledges the Russian Fund of the Fundamental Research for financial support through grant 13-02-00257. On of us, V.A.B., acknowledges also the Ministry of Education of the Russian Federation, grant 8412.

Contents

List of Figures

List of Notations

G — The Newton constant
e — electric charge
$\rho(\tau)$ — Shell radius as a function of the proper time τ
$M > 0$ — "bare" mass of the shell
$m_{\text{in}} > 0$ — mass of the internal metric
$m_{\text{out}} > 0$ — mass of the external metric
$\Delta m = m_{\text{out}} - m_{\text{in}}$ — see definition in (39)
$m_{\text{run}}(\rho)$ — running mass (84) and (85)
$m_{\text{in,run}}(\rho)$ — mass of the internal metric (84)
$m_{\text{in,run}}(\rho)$ — mass of the external metric (85)
$r_g = 2Gm_{\text{out}}/c^2 > 0$ — gravitational radius of the external metric
$r_\pm = Gm_{\text{in}} \pm \sqrt{G^2m_{\text{in}}^2 - Ge^2}$ — radii of horizons of the internal metric
$R_\pm$–region — space-time region with the signature $(+,-,-,-)$
$T_\pm$–region — space-time region with the signature $(-,+,-,-)$
$\sigma_{\text{in}}(\rho) = \pm 1$ and $\sigma_{\text{out}}(\rho) = \pm 1$ — sign-changing functions in (72)
ρ_0 — turning point (rejection) in (56)
$\rho_{0,\text{min}}$ — minimal turning point
$\rho_{0,-}$ — smaller turning point
$\rho_{0,+}$ — larger turning point
ρ_d — turning point (222) at the coincidence $\rho_{0,-} = \rho_{0,+}$
$\rho_{\sigma_{\text{in}}}$ — point of sign changing of σ_{in} in $T_\pm$–regions
$\rho_{\sigma_{\text{out}}}$ — point of sign changing of σ_{out} in $T_\pm$–regions (26)
$m_{\text{out,min}}$ — minimally possible mass m_{out} (197)
$m_{\text{in,min}}$ — minimally possible mass m_{in} (197)
Δm_{cr} — critical value Δm (181)
Δ — invariant function in (242)

i_0 — spatial infinities on the Carter–Penrose diagrams
$i_\pm$ — past and future temporal infinities
$\mathcal{J}^\pm$ — past and future null infinities
$R_\pm$–regions — 4D regions with signature $(+,-,-,-)$
$T_\pm$–regions — 4D regions with signature $(-,+,-,-)$
$\Delta m = m_{\text{out}} - m_{\text{in}}$ — defined in (242)
$r_g = 2Gm_{\text{out}} > 0$ — gravitational radius in the outer metrics
$r_\pm = Gm_{\text{in}} \pm \sqrt{G^2 m_{\text{in}}^2 - Ge^2}$ — horizon's radii in the inner metrics
$\rho(\tau)$ — shell's radius as a function of its proper time τ
ρ_0 — turning point at the shell trajectory
$\sigma_{\text{in}}(\rho) = \pm 1$ and $\sigma_{\text{out}}(\rho) = \pm 1$ — sign functions in (259) and (268)

Preface

It is elaborated the complete classification of the possible types of the spherically symmetric global geometries for two types of electrically charged shells: (1) The charged shell as a single source of the gravitational field, when internal space-time is flat, and external space-time is the Reissner–Nordström metric; (2) The neutralizing shell with an electric charge opposite to the charge of the internal source with the Reissner–Nordström metric and with the Schwarzschild metric outside the shell. Starting from Israel equations for the spherically symmetric thin shells we introduce the effective potential and show how it can be used in constructing, without further thorough investigation, the corresponding Carter–Penrose diagrams describing clearly the global geometry of the composite space-time manifolds. We demonstrate, how this new method works, by considering all possible configurations for the neutral thin dust shell immersed into different types of Reissner–Nordström electro-vacuum manifolds.

Chapter 1

Introduction

The structure of any spherically symmetric space-time is completely determined by two invariant functions of two variables. Indeed, locally, the general spherically symmetric metric can be written as

$$ds^2 = A^2 dt^2 + 2H dt dq - B^2 dq^2 - R^2 d\sigma^2 , \tag{1}$$

where $A(t,q), H(t,q)$ and $B(t,q)$ are functions of the time coordinate, t, and some radial coordinate, q, $d\sigma^2$ is the line element of a $2-dim$ unit sphere, and $R(t,q)$ is the radius of this sphere in the sense that its area equals $4\pi R^2$. Therefore, we are, actually, dealing with the invariant function $R(t,q)$ and the two-dimensional metric, which by suitable coordinate transformation can always be put in the conformally flat form

$$ds_2^2 = \gamma_{ik} dx^i dx^k = \omega^2(t,q)(dt^2 - dq^2) , \quad i,k = 0,1 . \tag{2}$$

This proves the above statement about two functions of two variables.

The first invariant function is, of course, the radius $R(t,q)$. By geometrical reasons, we choose for the second function the invariant (notations are obvious)

$$\Delta = \gamma^{ik} \frac{\partial R}{\partial x^i} \frac{\partial R}{\partial x^k} = \frac{1}{\omega^2} \left(\dot{R}^2 - R'^2 \right) . \tag{3}$$

This is nothing more but the square of the normal vector to the surfaces of constant radii, $R(t,q) = const$. The invariant function Δ brings a very important geometrical information. If $\Delta < 0$, the surfaces $R = const$ are time-like, such regions are called the $R_\pm$–regions, the signs "$\pm$" being denote the sign of a spatial derivative of the radial function R. If $\Delta > 0$, the regions are called the $T_\pm$–regions, depending on the sign of the corresponding time derivative (inevitable expansion or inevitable contraction), and the surfaces $R = const$ are space-like. The $R_\pm-$ and $T_\pm-$ regions are separated by the apparent horizons with $\Delta = 0$. It is the set of these regions and horizons together with the boundaries (infinities and that determines the global geometry. The boundaries are to be chosen in such a way that the space-time becomes geodesically complete, namely, all the time-like and null geodesics should start and end either at infinities or at singularities.

The causal structure of geodesically complete spherically symmetric space-times can be best seen on the conformal Carter–Penrose diagrams, where each point represents a sphere, and infinities are brought to the final distances. Since every 2-dimensional space-time is (locally) conformally flat, its Carter–Penrose diagram is the set of that for the $2-dim$ Minkowski manifold. To see how the latter looks like, let us, first, transform the Minkowski metric $ds^2 = dt^2 - dx^2$ to the double-null coordinates $u = t - x$ (retarded time) and $v = t + x$ (advanced time), then $ds^2 = dudv$. We will use the convention that on the diagram the time coordinate increases from down to up, the spatial coordinate - from left to right, and the null curves ($u = const$, $v = const$ are the straight lines with the slope $\pm 45°$. Making one more transformation

$$\begin{aligned} u' &= \arctan u\,, \quad -\frac{\pi}{2} \le u' \le \frac{\pi}{2} \\ v' &= \arctan v\,, \quad -\frac{\pi}{2} \le v' \le \frac{\pi}{2} \end{aligned} \tag{4}$$

one gets

$$\begin{aligned} ds^2 &= \Omega^2 ds'^2\,, \quad \Omega = \frac{1}{\cos u' \cos v'} \\ ds'^2 &= du'dv' = dt'^2 - dx'^2\,. \end{aligned} \tag{5}$$

Formally, the metric ds'^2 looks exactly as the starting one, but now coordinates (u', v') and (t', x') run the finite intervals.

The Carter–Penrose diagram for the complete $2-dim$ Minkowski space-time $(-\infty < t < \infty,\ -\infty < x < \infty)$ is shown in Fig. 1.

Here $J^\pm(J'^\pm)$ are null future $(v'(u') = \frac{\pi}{2}, v(u) = \infty)$ and past $(u'(v') = -\frac{\pi}{2}, u(v) = -\infty)$ infinities, $i_\pm$ are future and past $(t' = \pm\frac{\pi}{2})$ temporal infinities, and $i_0(i'_0)$ are spatial $(x' = \pm\frac{\pi}{2}, x = \pm\infty)$ infinities. If the corresponding

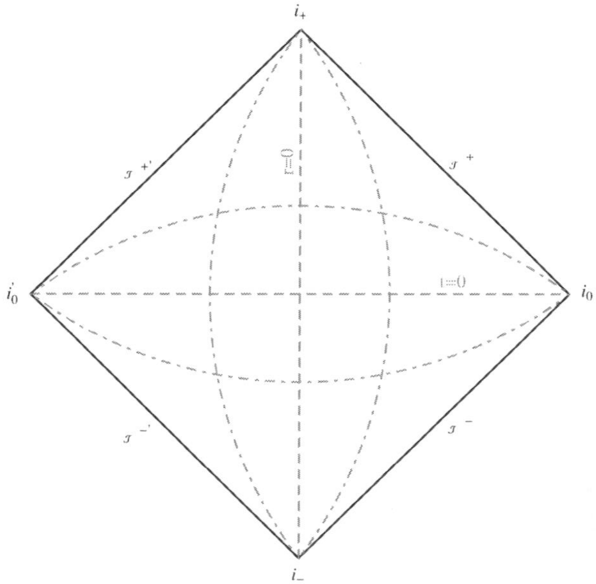

Figure 1. The Carter–Penrose diagram for the complete $2-dim$ Minkowski space-time $(-\infty < t < \infty,\ -\infty < x < \infty)$. The horizontal dashed curves represent $t = const$ lines, while the vertical ones are for $x = const$.

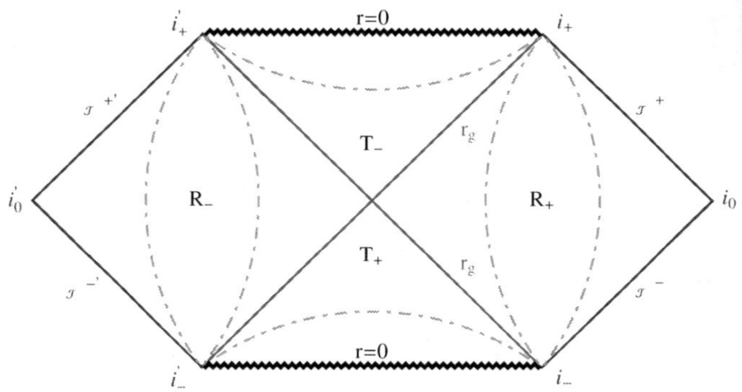

Figure 2. The complete Carter–Penrose diagram of the Schwarzschild metric.

conformally flat metric is not complete in the sense that one of the coordinates starts from or ends at the finite boundary value (like, for example, the zero radius value in the case of spherical symmetry), then one should cut the above square along the corresponding diagonal (in general, along some time-like os space-like curve), and such part of the complete Carter–Penrose diagram will be a triangle with the vertical (left for R_+–regions and right for R_-–regions) or horizontal (for $T_\pm$–regions) boundary.

Chapter 2

Conformal Carter–Penrose Diagrams

Both the Schwarzschild and Reissner–Nordström metrics look the same in the so-called curvature coordinates:

$$ds^2 = F dt^2 - \frac{1}{F} dR^2 - R^2 (d\vartheta^2 + \sin^2 \vartheta d\varphi^2) , \tag{6}$$

where R - radius $(0 \leq R < \infty), F = F(R)$, and ϑ and φ are spherical angles. The two-dimensional part can easily be written in the conformally flat form by introducing the "tortoise" coordinate $R^\star$:

$$\begin{aligned} dR^\star &= \frac{dR}{|F|} , \\ ds_2^2 &= F \left(d\xi^2 - dR^{\star 2} \right) . \end{aligned} \tag{7}$$

In the $R_\pm$–regions $F = -\Delta > 0$ and $R^\star$ plays the role of the spatial (radial) coordinate q, while ξ is the time coordinate t. In the $T_\pm$–regions, $R^\star$ plays the role of the time coordinate t, while ξ is the spatial coordinate q.

Consider, first, the Schwarzschild metric. In this case

$$F = 1 - \frac{2Gm}{R} , \tag{8}$$

where G is the Newton's gravitational constant m is the total mass of the gravitating system measured by distant observers (at infinity), and we put the speed of light $c = 1$. For $R > r_g = 2Gm$ we have the the R–region, and for $R < r_g$ - the T–region. The event horizon coincides with the apparent horizon at $R = r_g$ (gravitational, or Schwarzschild, radius). At $R = 0$ we encounter the (space-like) curvature singularity. The complete Carter–Penrose diagram looks as follows in Fig. 2.

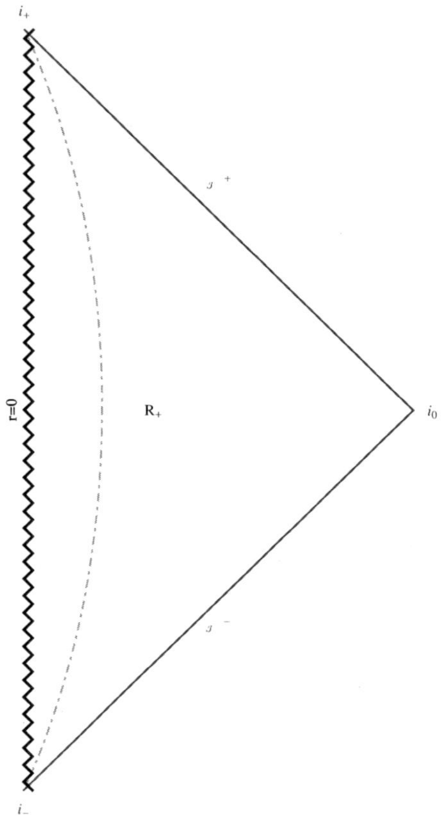

Figure 3. The complete Carter–Penrose diagram of the Reissner–Nordström naked singularity, $Gm^2 < e^2$.

(3) $Gm^2 < e^2$ - no black hole, naked singularity at $R - 0$. The Carter–Penrose diagram is very simple (see Fig. 3. There are two isometric $R_\pm$–regions bounded by two apparent (past and future) horizons at $R = r_g$ and two asymptot-

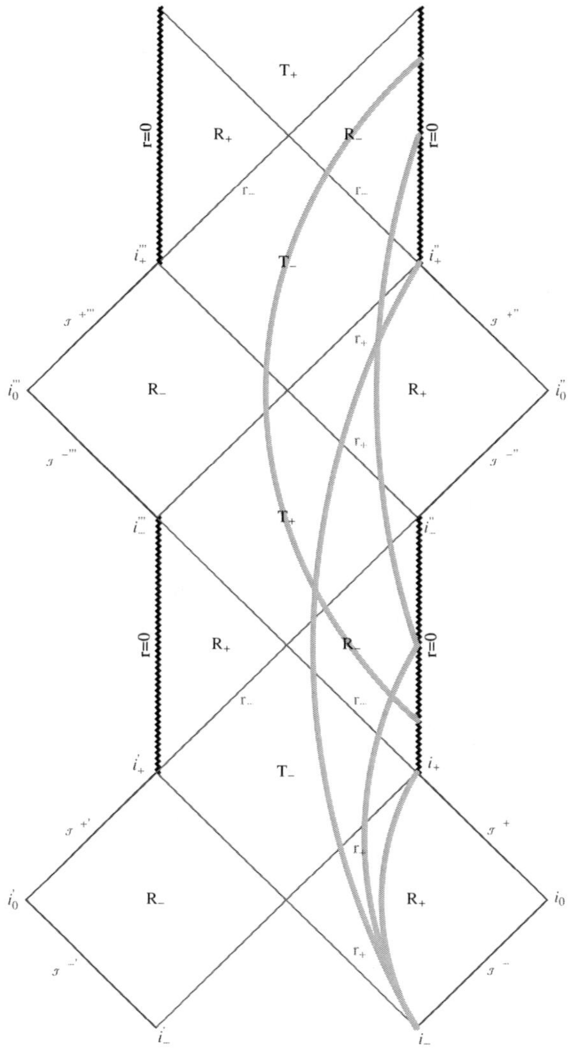

Figure 4. The Carter–Penrose diagram for Reissner–Nordström black hole. Thick curves are the trajectories of the shell with turning points.

ically flat regions with corresponding future and past temporal $(i_{\pm}, i'_{\pm})$, future and past null $(J_{\pm}, J'_{\pm})$ and spatial (i_0, i'_0) infinities. Also we have two T–regions

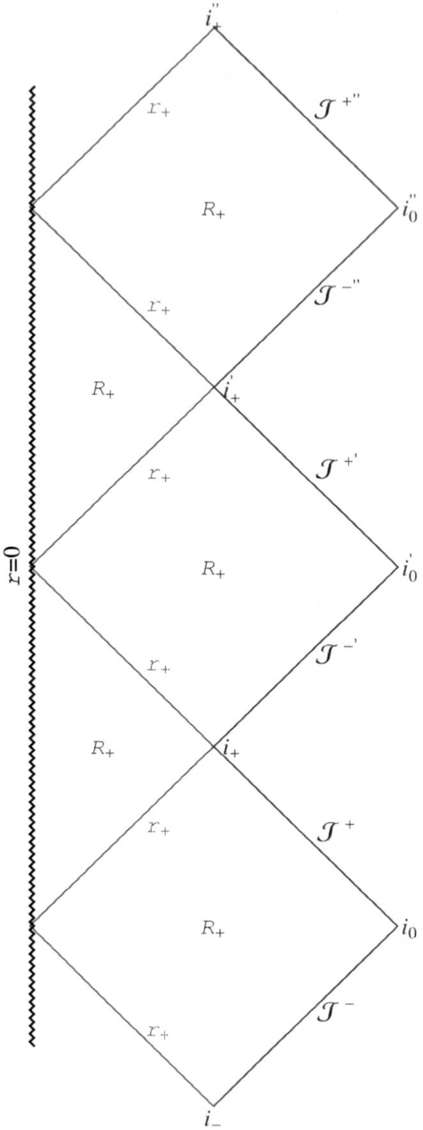

Figure 5. Extreme Reissner–Nordström black hole, $Gm^2 = e^2$.

(T_+ and T_-) bounded by the apparent horizons at $R = r_g$ and future and past space-like singularities at $R = 0$. This is called the eternal Schwarzschild black hole. The gravitational source is concentrated on these two space-like singularities, i.e., it exists only for one moment in the past and reappears again for one moment in the future.

The causal structure of the Reissner–Nordström space-time is much more complex. The function F equals now

$$F = 1 - \frac{2Gm}{R} + \frac{Ge^2}{R^2}, \tag{9}$$

e is the electric charge. There are three different cases

(1) $Gm^2 > e^2$ — Reissner–Nordström black hole, equation $F = 0$ has two nonequal real roots $r_\pm$,

$$r_\pm = Gm \pm \sqrt{G^2m^2 - Ge^2}. \tag{10}$$

According to the signs of F, we have the R–regions for $r_+ < R < \infty$ and $0 \leq R < r_-$, T–regions in-between, $r_- < R < r_+$, and two apparent horizons at $R = r_\pm$, the external one, r_+, playing the role of the event horizon, and the inner, r_-, - the Cauchy horizon. The geodesically complete Carter–Penrose diagram is the ladder extended infinitely to the past and to the future as shown in Fig. 4.

In the complete (eternal) Reissner–Nordström black hole space-time both the the gravitational source and the electric charge(s) are concentrated on two (for each part of the ladder) time-like singularities $R = 0$ (left and right on the diagram), the signs of the electric charges on them being opposite.

(2) $Gm^2 = e^2$ — extreme Reissner–Nordström black hole. Equation $F = 0$ has the double root $r_+ = r_- = Gm = \sqrt{G}|e|$. We have R–regions everywhere except the apparent (event) horizon at $R = r_+ = r_-$, as shown in Fig. 5.

Chapter 3

Thin Shell Formalism

The thin shell is a hyper-surface in the space-time on which the energy-momentum tensor is singular. If such a hyper-surface is time- or space-like, one can introduce in its vicinity the so-called Gaussian normal coordinates, and the line element can be written as

$$ds^2 = \varepsilon dn^2 + \gamma_{ij}(n,x)dx^i dx^j\,, \tag{11}$$

n is the coordinate in the normal direction to the shell, and x^i - coordinates on the shell, $\varepsilon = +1$ in the space-like case and $\varepsilon = -1$ in the time-like case. The surface is supposed to be located at $n = 0$. The energy-momentum tensor T^μ_ν is proportional to δ-function,

$$T^\mu_\nu = S^\mu_\nu\,\delta(n)\,, \tag{12}$$

S^μ_ν is called the surface energy-momentum tensor. The dynamics of the thin shell is governed by the Israel equations [1] (see also [3] – [8]), obtained by integrating the Einstein equations across the shell. First of all, one gets $S^n_n = S^i_n = 0$, this can be considered as the definition of the thin shell. The Israel equations are

$$\varepsilon\left([K_{ij}] - \gamma_{ij}\,[K]\right) = 8\pi G S_{ij}\,, \tag{13}$$

supplemented by the Bianchi identity for the shell

$$S^j_{i|j} + [T^n_i] = 0\,. \tag{14}$$

Here $K_{ij} = -(1/2)\partial\gamma_{ij}/\partial n$ is the extrinsic curvature tensor, K is its trace, brackets $[\,] = (out) - (in)$ is the jump across the shell, the vertical line denotes the covariant derivative with respect to the metric γ_{ij}. In what follows we will be dealing with the time-like shells only, so, $\varepsilon = -1$.

In the case of spherical symmetry everything is simplified drastically. The metric becomes

$$ds^2 = -dn^2 + \gamma_{00}(n,\tau)d\tau^2 - \rho^2(n,\tau)d\sigma^2, \tag{15}$$

$\rho(0,\tau)$ is the shell radius as a function of the proper time τ of the observer, sitting on this shell, $n < 0$ inside and $n > 0$ outside. The mixed components of the surface energy momentum tensors are S_0^0 (surface energy density) and $S_2^2 = S_3^3$ (surface tension), and the Israel equations reduced to one constraint and one dynamical equations, namely,

$$\begin{aligned} \left[K_2^2\right] &= 4\pi G S_0^0 \\ \left[K_0^0\right] + \left[K_2^2\right] &= 8\pi G S_2^2 . \end{aligned} \tag{16}$$

The supplement equation is now

$$\dot{S}_0^0 + \frac{2\dot{\rho}}{\rho}\left(S_0^0 - S_2^2\right) + [T_0^n] = 0 . \tag{17}$$

We are interested in the situation when both inside and outside the shell the space-time is (electro)-vacuum one, hence, $T_0^n = 0$. For the sake of simplicity we will consider the dust shell, for which $S_2^2 = 0$. Then,

$$S_0^0 = \frac{M}{4\pi\rho^2}, \tag{18}$$

where $M = const$ is the "bare" mass of the shell (without the gravitational mass defect). Thus, we need only the first, constraint, equation. In order to go further we have to calculate

$$K_2^2 = -\frac{1}{\rho^2}K_{22} = -\frac{1}{2\rho^2}\frac{\partial(\rho^2)}{\partial n} = -\frac{\rho_{,n}}{\rho} . \tag{19}$$

But, from definition of the invariant Δ it follows

$$\begin{aligned} \Delta &= \dot{\rho}^2 - \rho_{,n}^2 \\ \rho_{,n} &= \sigma\sqrt{\dot{\rho}^2 - \Delta} \\ K_2^2 &= -\frac{\sigma}{\rho}\sqrt{\dot{\rho}^2 - \Delta} . \end{aligned} \tag{20}$$

Here $\sigma = \pm 1$ depending on whether radii increasee ($\sigma = +1$) in the outward normal direction or decrease ($\sigma = -1$). Thus, the sign of σ coincides with that of the R–region, and it can change only in the T–regions. Finally, the only equation we will need in our analysis is

$$\sigma_{in}\sqrt{\dot{\rho}^2 - \Delta_{in}} - \sigma_{out}\sqrt{\dot{\rho}^2 - \Delta_{out}} = \frac{GM}{\rho}. \tag{21}$$

Since in our case $\Delta = -F$, we have

$$\sigma_{in}\sqrt{\dot{\rho}^2 + 1 - \frac{2Gm_{in}}{\rho} + \frac{Ge_{in}^2}{\rho^2}} - \sigma_{out}\sqrt{\dot{\rho}^2 + 1 - \frac{2Gm_{out}}{\rho} + \frac{Ge_{out}^2}{\rho^2}} = \frac{GM}{\rho}. \tag{22}$$

We will not consider exotic matter shells, so $M > 0$. From the above constraint equation (that is nothing more but the energy conservation law) it follows that for the qualitative analysis one needs to investigate the behavior of the function $\rho(\tau)$ only at several special points: $\rho \to \infty$, $\dot{\rho} = 0$, $\rho = 0$ and $\rho = \rho_\sigma$ where $\sigma_{out}(\sigma_{in})$ changes its sign. Examples for using the thin shell formalism see, e. g., in [8] – [9].

Chapter 4

Charged Shell with Minkowski Space-time Inside

We start by considering the case of a thin charged shell with the Minkowski space-time (containing the world-line $r = 0$) inside and Reissner–Nordström one outside . This means that $m_{\text{in}} = e_{\text{in}} = 0$, $F_{\text{in}} = 1$, $\sigma_{\text{in}} = +1$ and the "naked" mass of the shell $M = const$. The equation for shell dynamics (22) in this particular case takes the form

$$\sqrt{\dot{\rho}^2 + 1} - \sigma_{\text{out}} \sqrt{\dot{\rho}^2 - \frac{2Gm}{\rho} + \frac{Ge^2}{\rho^2}} = \frac{GM}{\rho}, \tag{23}$$

where it is written $m = m_{\text{out}}$ for brevity. By squaring equation (23) we get

$$m = M\sqrt{\dot{\rho}^2 + 1} - \frac{2GM - e^2}{2\rho}. \tag{24}$$

and

$$\sigma_{\text{out}} \sqrt{\dot{\rho}^2 + 1 - \frac{2Gm}{\rho} + \frac{Ge^2}{\rho^2}} = \frac{m}{M} - \frac{GM^2 + e^2}{2M\rho}. \tag{25}$$

It easily seen from (25) that $m > M$ for infinite (unbound) motions and $m < M$ for finite (bound) ones. From (25) we obtain the relations which define the sign of σ_{out}:

$$\sigma_{\text{out}} = \text{sign}\left(\frac{m}{M} - \frac{GM^2 + e^2}{2M\rho}\right). \tag{26}$$

and determine the radius ρ_σ, where $\sigma_{\rm out}$ changes its sign:

$$\rho_\sigma = \frac{GM^2 + e^2}{2m}. \tag{27}$$

It happens when $\dot\rho^2 + F_{\rm out} = 0$ where $F_{\rm out} < 0$, so, it may take place only in T–region, i.e., only in the case $e^2 < Gm^2$, the Reissner–Nordström black hole metric. Formally, the value for ρ_σ can always be calculated. But it lies on the shell trajectory, only if

$$r_- < \rho_\sigma < r_+, \tag{28}$$

This is equivalent (as can be easily verified) to

$$r_- < GM < r_+. \tag{29}$$

For infinite motions $GM < Gm < r_+$, so, the value of σ remains unchanged during the shell evolution, if $GM < r_- = Gm - \sqrt{G^2m^2 - Ge^2}$, and this is equivalent to

$$\frac{m}{M} < \frac{1}{2} + \frac{e^2}{2GM^2}. \tag{30}$$

In the case of finite motions $r_- < Gm < GM$, so, the value of σ remains unchanged if $GM > r_+ = Gm + \sqrt{G^2m^2 - Ge^2}$, and we obtain the same relation for the total mass m as before. Thus, we have the universal relations for the shell motion in the case when the external solutions are the Reissner–Nordström black holes with $Gm^2 > e^2$: σ changed its sign on the shell trajectory if

$$\frac{m}{M} > \frac{1}{2} + \frac{e^2}{2GM^2} \tag{31}$$

and it remains unchanged if

$$\frac{m}{M} < \frac{1}{2} + \frac{e^2}{2GM^2} \tag{32}$$

Now, let us define the relation for the turning point $\rho = \rho_0$, where $\dot\rho = 0$. The turning points can be situated only in the $R_\pm$–regions. By putting $\dot\rho = 0$ in (24) we find that it may be no turning points at all or only one turning point:

$$\rho_0 = \frac{GM^2 - e^2}{2(M - m)}. \tag{33}$$

Evidently, the turning point exists if $\text{sign}(M-m) = \text{sign}(GM^2 - e^2)$, i.e., for infinite motions $(m > M)$ it happens for $e^2 > GM^2$, while for finite motions $(m < M)$ the turning point exists by definitions, so the finite motion itself is possible, only if $e^2 < GM^2$.

It is clear from (23) that the shell evolution is completely determined by the values of two parameters, bare mass M and electric charge e that characterizes the shell itself, and the total mass of the system, m, depending on the initial conditions which includes the initial value of σ_{out}. For infinite motions it is natural to put the initial conditions at infinity, $\rho \to \infty$. Since we adopt the physically acceptable signs for the masses, $M > 0$, $m > 0$, we find that $\sigma_{out}(\infty) = +1$, and the value of the total mass m is determined by the initial velocity of the shell there. For finite motions at $\rho \to 0$, $\sigma(0) = -1$, and the value of the total mass is determined by the value of the turning point ρ_0 and the value of σ there. It should be noted here that, given the shell parameters and initial values for $\rho(0)$ and $\dot{\rho}(0)$, we are not able to continue unambiguously the solutions beyond the Cauchy horizons present in the complete Reissner–Nordström black hole space-times. To avoid any inconvenience we decided to do this using equations of motion for our shells and demanding $\rho(\tau)$ and $\dot{\rho}(\tau)$ to be continuous functions.

We start our classification by specifying the relation between the shell parameters, M and e, and then determine the trajectories of the shells and corresponding global geometries depending on the values of the total mass m. Evidently, we should distinguish two types of the shells, that ones with $GM^2 < e^2$ and with $e^2 < GM^2$.

4.1. Type I: $GM^2 < e^2$

There are three different cases, depending on where we insert the total mass m into the above inequality: to the left, in the middle, or to the right.

I-1: $Gm^2 < GM^2 < e^2$

Finite motion with no turning point, what is impossible.

I-2: $GM^2 < Gm^2 < e^2$

Infinite motion with a turning point. Outside we have the Reissner–Nordström metric with naked singularity. But, since $\rho_0 < r < \infty$, this singularity is "hid-

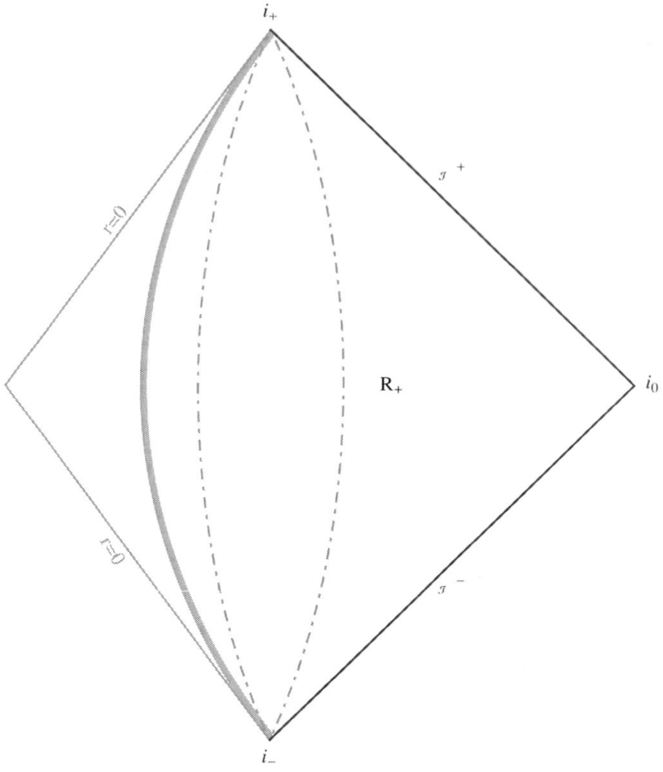

Figure 6. Case I-2: Infinite motion with a turning point, $GM^2 < Gm^2 < e^2$.

den" inside the shell where, instead, the space-time is flat. The Carter–Penrose diagram looks as follows in Fig. 6.

I-3: $GM^2 < e^2 < Gm^2$

Infinite motion with turning point. The exterior metric is the Reissner–Nordström black hole with two horizons at $r_\pm = Gm \pm \sqrt{G^2m^2 - Ge^2}$. A turning point should lie in R–region, but it could happen outside the event horizon, $\rho_0 > r_+$, as well as inside the inner horizon, $\rho_0 < r_-$. It is not difficult to show that the first of the inequalities is equivalent to $(r_+ - GM)^2 < 0$ what is impossible. Since for $0 < r < r_-$ there exist two R–regions, R_+ and R_-, and at infinity

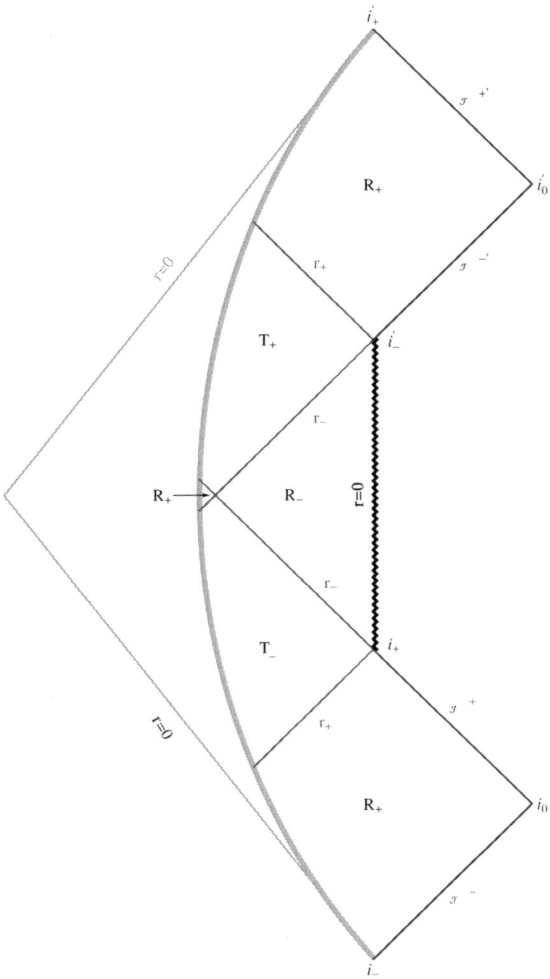

Figure 7. Subcase I-3(a): On the shell trajectory σ_{out} does not change its sign.

$\sigma_{out} = +1$, we have, accordingly, two subcases.

Subcase I-3(a):

On the shell trajectory σ_{out} does not change its sign, consequently,

$$\frac{m}{M} < \frac{1}{2} + \frac{e^2}{2GM^2}, \tag{34}$$

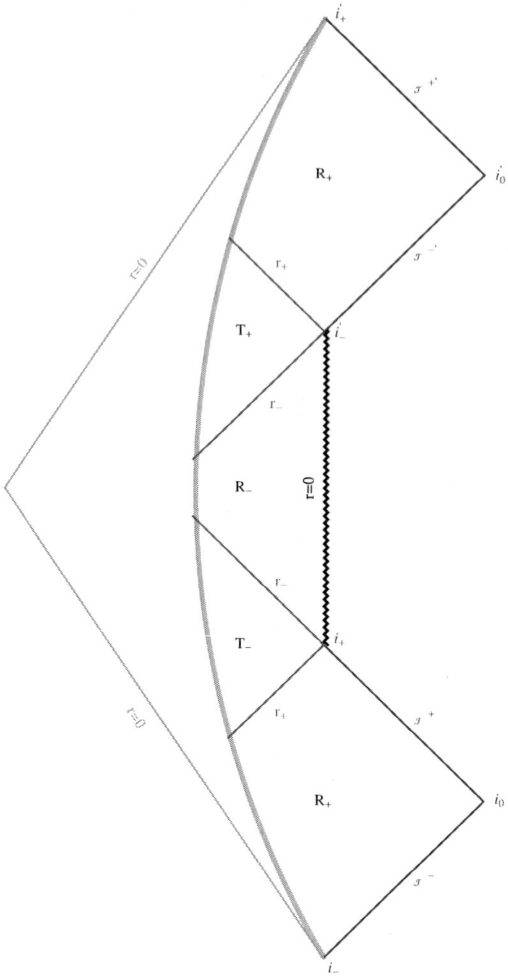

Figure 8. Subcase I - 3(b): σ_{out} changes its sign, and the turning point lies in R_-–region, $0 < \rho_0 < r_-$.

the turning point lies in R_+–region, $0 < \rho_0 < r_-$, and the Carter–Penrose diagram looks as follows in Fig. 7.

Subcase I-3(b):

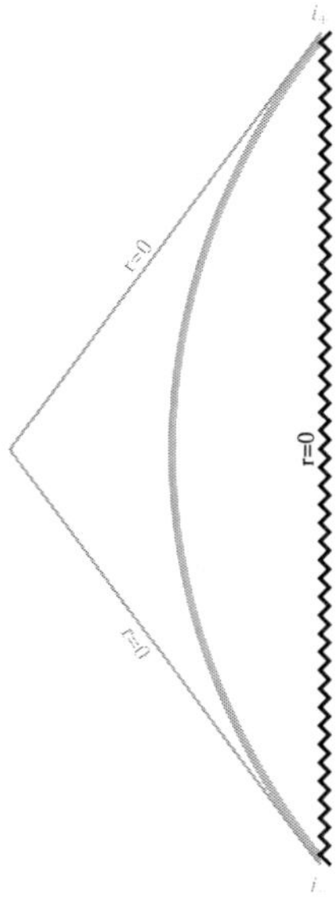

Figure 9. Subcase II-1: Finite motion with turning point, $Gm^2 < e^2 < GM^2$.

σ_{out} changes its sign, and the turning point lies in R_-–region, $0 < \rho_0 < r_-$,

$$\frac{m}{M} > \frac{1}{2} + \frac{e^2}{2GM^2}, \tag{35}$$

The conformal diagram is shown in Fig. 8.

4.2. Type II: $e^2 < GM^2$

Again, we have three different cases depending on the values of total mass m.

II-1: $Gm^2 < e^2 < GM^2$

Finite motion with turning point. Outside is the Reissner–Nordström geometry with naked singularity, and $\sigma_{out} = -1$ everywhere. The conformal diagram is very simple and shown in Fig. 9. This diagram is rather curious. The difference

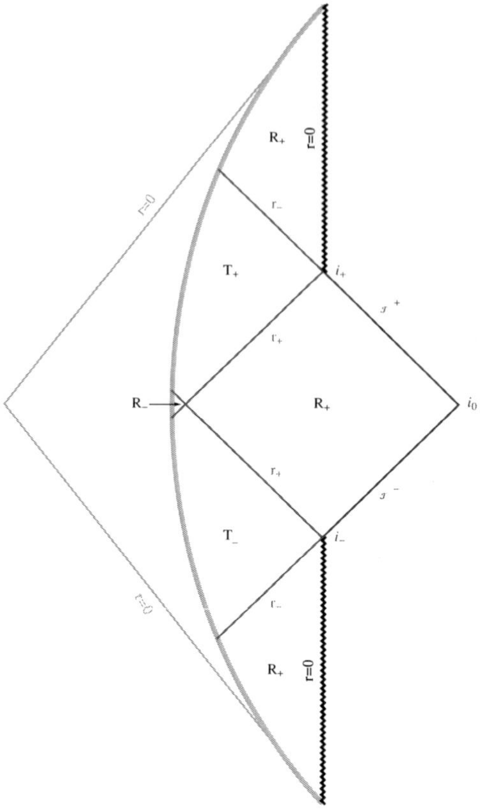

Figure 10. Subcase II-2(a): The turning point lies in the asymptotically flat R_-–region on the other side of the Einstein–Rosen bridge.

between the left and right parts is only in the nature of the world lines $r = 0$, the first one is nonsingular, while the other does, and it is there the electric charge with the opposite sign to that of the shell's is concentrated. By dashed curves we show the curves of constant radii, $0 < r < \rho_0$.

II-2: $e^2 < Gm^2 < GM^2$

Finite motion, therefore, $0 \le \rho \le \rho_0$. Again, we have two possibilities depending on whether σ_{out} changes its sign on the trajectory, or not. But now trajecto-

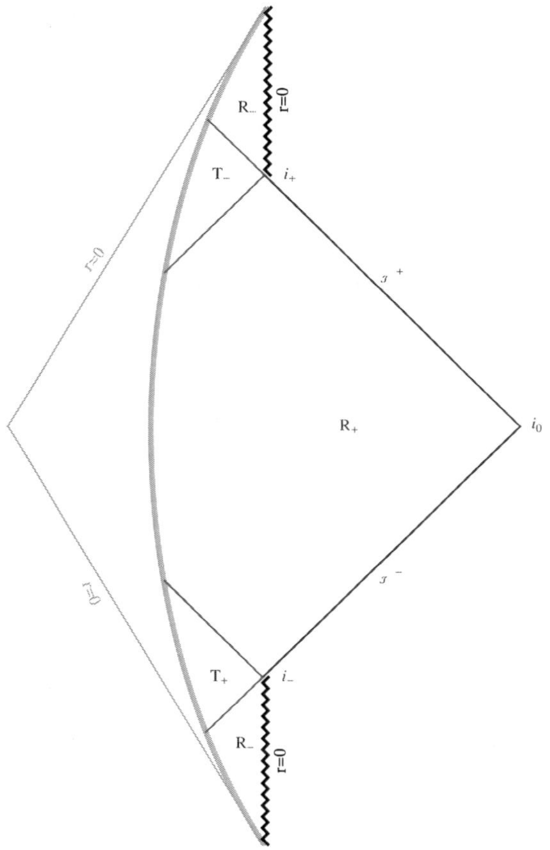

Figure 11. Subcase II-2(b): σ_{out} changes its sign during the shell evolution, and the turning point lies in the asymptotically flat R_+–region on "our" side of the Einstein–Rosen bridge.

ries start at $\rho = 0$ where $\sigma_{out} = -1$, and, as can be shown, the turning point ρ_0 lies outside the event horizon, $\rho_0 > r_+$. Thus,

Subcase II-2(a):

$$\frac{m}{M} < \frac{1}{2} + \frac{e^2}{2GM^2}\,. \tag{36}$$

Everywhere on the trajectory $\sigma_{out} = -1$, so, the turning point lies in the asymptotically flat R_-–region on the other side of the Einstein–Rosen bridge. The

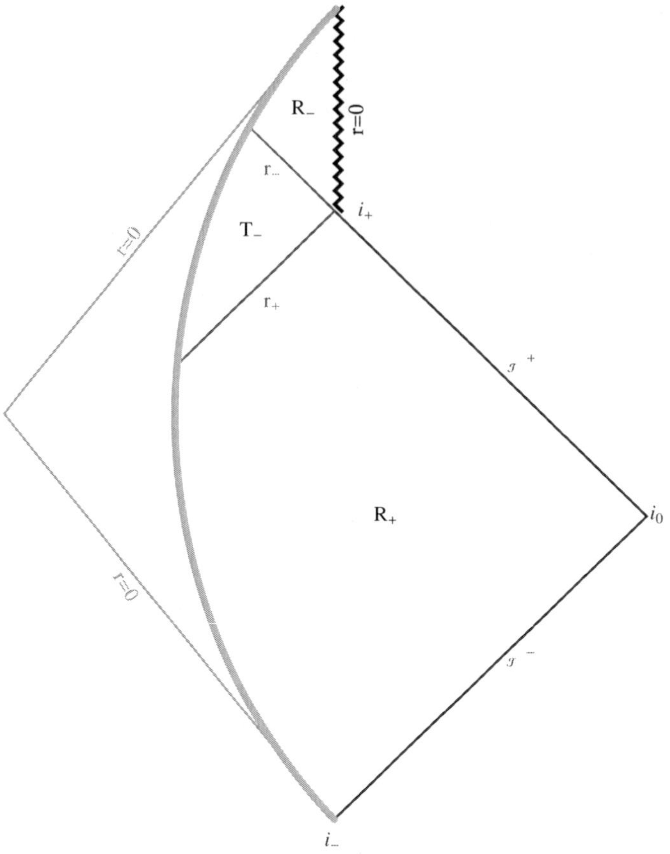

Figure 12. Subcase II-3: Infinite motion with no turning point, $e^2 < GM^2 < Gm^2$.

Carter–Penrose diagram looks as follows in Fig. 11.

Subcase II-2(b):

$$\frac{m}{M} > \frac{1}{2} + \frac{e^2}{2GM^2}, \tag{37}$$

σ_{out} changes its sign during the shell evolution, and the turning point lies in the asymptotically flat R_+–region on "our" side of the Einstein–Rosen bridge. The conformal diagram is shown in Fig. 10.

II-3: $e^2 < GM^2 < Gm^2$

Infinite motion with no turning point. It starts at infinity in R_+–region with $\sigma_{out} = +1$ and ends at $\rho = 0$ with $\sigma_{out} = -1$. Of course, there exists also the reverse motion. Both of them are shown on Carter–Penrose diagram below in Fig. 12.

Chapter 5

Neutralizing Shell

Now we consider the charged shell with the electric charge opposite in sign to the charge of the internal Reissner–Nordström metric with the mass $m_{\rm in}$ and electric charge e. The shell has a bare mass M and charge $-e$, which neutralizes the charge of the internal source. As a result, the external metric is the Schwarzschild one with a mass $m_{\rm tot} = m$.

Now equation for shell dynamics (22) can be written as

$$\sigma_{\rm in}\sqrt{\dot{\rho}^2 + 1 - \frac{2Gm_{\rm in}}{\rho} + \frac{Ge^2}{\rho^2}} - \sigma_{\rm out}\sqrt{\dot{\rho}^2 + 1 - \frac{2Gm}{\rho}} = \frac{GM}{\rho}, \tag{38}$$

where the total mass $m_{\rm tot} = m_{\rm out} = m$.

Now both $\sigma_{\rm in}$ and $\sigma_{\rm out}$ can change sign during shell motion.

Changing Sign in $\sigma_{\rm out}$

By squaring (38) to exclude $\sigma_{\rm in}$ we obtain

$$m - m_{\rm in} = \Delta m = M\sigma_{\rm out}\sqrt{\dot{\rho}^2 + 1 - \frac{2Gm}{\rho}} + \frac{GM^2 - e^2}{2\rho}, \tag{39}$$

and so

$$\sigma_{\rm out} = \text{sign}\left(\Delta m - \frac{GM^2 - e^2}{2\rho}\right). \tag{40}$$

In the limiting cases we have $\sigma_{\rm out}(\infty) = \Delta m$ and $\sigma_{\rm out}(0) = -{\rm sign}(GM^2 - e^2)$. From (40) we find now the radius of changing sign for $\sigma_{\rm out}$:

$$\rho_\sigma = \frac{GM^2 - e^2}{2\Delta m}. \tag{41}$$

The radius ρ_σ can change sign only in the T–regions of the Schwarzschild metric, i. e.

$$0 < \rho_\sigma < 2Gm. \tag{42}$$

From the left inequality in (42) it follows that ${\rm sign}[(GM^2 - e^2)/\Delta m] = +1$ only if ${\rm sign}(GM^2 - e^2) = {\rm sign}(\Delta m)$, and so $\sigma_{\rm out}(0) = -\sigma_{\rm out}(\infty)$. From the right inequality in (42) it follows that

$$\frac{GM^2 - e^2}{\Delta m} < 4Gm. \tag{43}$$

If $\sigma_{\rm out}(\infty) + 1$, then we have relations $\Delta m > 0$ and $GM^2 - e^2 > 0$. Inequality (42) now can be written as

$$GM^2 - e^2 < 4Gm\Delta m = 4Gm(\Delta m)^2 + 4Gm_{\rm in}\Delta m. \tag{44}$$

or

$$4Gm(\Delta m)^2 + 4Gm_{\rm in}\Delta m - (GM^2 - e^2) > 0. \tag{45}$$

For the r.h.s. there are one negative and one positive root. For a positive root we obtain:

$$\Delta m_\sigma = \frac{m_{\rm in}}{2}\left(\sqrt{1 + \frac{GM^2 - e^2}{Gm_{\rm in}^2}} - 1\right). \tag{46}$$

It can be seen that $\Delta m > \Delta m_\sigma$.

In the opposite case $\sigma_{\rm out}(\infty) = +1$ we have relations $\Delta m < 0$ and so $GM^2 - e^2 < 0$. Inequality (42) now can be written as $GM^2 - e^2 > 4Gm\Delta m$ or

$$4Gm(\Delta m)^2 + 4Gm_{\rm in}\Delta m + (e^2 - GM^2) < 0. \tag{47}$$

Now for the r.h.s. there are two negative roots or the roots are absent at all:

$$\Delta m_{\rm cr} = \frac{m_{\rm in}}{2}\left(-1 \pm \sqrt{1 - \frac{e^2 - GM^2}{Gm_{\rm in}^2}}\right). \tag{48}$$

From (47) we see that condition for the existence of two roots is

$$Gm_{\rm in}^2 > e^2 - GM^2. \tag{49}$$

Now we consider in details the behavior of $\sigma_{\rm in}$

Changing Sign in σ_{in}

Analogously, by squaring (38) to exclude σ_{out} we obtain

$$M\sigma_{\text{in}}\sqrt{\dot{\rho}^2+1-\frac{2Gm_{\text{in}}}{\rho}+\frac{Ge^2}{\rho^2}}=\Delta m+\frac{GM^2+e^2}{2\rho}, \tag{50}$$

and so

$$\sigma_{\text{in}}=\text{sign}\left(\Delta m+\frac{GM^2+e^2}{2\rho}\right). \tag{51}$$

In the limiting cases we have $\sigma_{\text{in}}(\infty)=\text{sign}(\Delta m)=\sigma_{\text{out}}(\infty)$ and $\sigma_{\text{in}}(0)=+1$. From (51) we find now the radius of changing sign for σ_{in}:

$$\rho_\sigma(\text{in})=-\frac{GM^2+e^2}{2\Delta m}. \tag{52}$$

Now there are possible the both cases, $\Delta m \gtrless 0$.

If $\Delta m>0$, then $\sigma_{\text{in}}=+1$ everywhere. In particular, $\sigma_{\text{in}}(0)=+1$ at $\rho\to 0$.

The change of sign in $\sigma_{\text{in}}=+1$ is possible only if $\Delta m<0$. This point must be in the T–region of the Reissner–Nordström metric, and so

$$r_- < \rho_\sigma(\text{in}) < r_+, \tag{53}$$

where $r_\pm = Gm_{\text{in}}\pm\sqrt{G^2m_{\text{in}}^2-Ge^2}$. Inequality (51) may be written also as

$$r_- < \rho_\sigma(\text{in}) < r_+, \tag{54}$$

These relations may rewritten as

$$r_- < \frac{e^2(\Delta m)}{GM^2m_{\text{in}}} < r_+, \tag{55}$$

5.1. Turning Points $\dot{\rho}^2=0$

From (39) we have an equation for the turning point $\rho=\rho_0$:

$$\left(\Delta m-\frac{GM^2-e^2}{2\rho_0}\right)^2=M\left(1-\frac{2Gm}{\rho_0}\right) \tag{56}$$

By introducing a new variable $x = (2M\rho_0)^{-1}$, we write equation $\dot{\rho}^2 = 0$ as

$$(GM^2 - e^2)^2 x^2 + 2x\left[\frac{\Delta m}{M}e^2 + GM(m + m_{\rm in})\right] + \frac{(\Delta m)^2}{M^2} - 1 = 0 \tag{57}$$

with roots

$$x_\pm = \frac{1}{(GM^2 - e^2)^2}\left\{-B \pm \sqrt{B^2 - (GM^2 - e^2)^2\left[\frac{(\Delta m)^2}{M^2} - 1\right]}\right\}, \tag{58}$$

where

$$B = \left[\frac{\Delta m}{M}e^2 + GM(m + m_{\rm in})\right]. \tag{59}$$

Finite motion — there are always two real root. Infinite motion — there are no real roots at all or there are two negative roots.

5.2. Infinite Motion with $\rho \to \infty$

We start our consideration with the case of infinite motion, when $\rho \to \infty$. In infinite motion $\sigma_{\rm out}(\infty) = \pm 1$ (R_+–region). From $\Delta m = M\sqrt{\dot{\rho}^2 + 1}$ at $\rho \to \infty$ it follows that $\Delta m > M > 0$. From (41) it follows that $\rho_\sigma(in) < 0$ and does not exist. Therefore, $\sigma_{\rm in} = +1$ for infinite motion. See Fig. 13.

The Case $\sigma_{\rm out}(\infty) = {\rm sign}(\Delta m) = +1$

In this case $\Delta m > 0$ and so $\sigma_{\rm in} = 1$ everywhere, $\Delta m > M$ and $\sigma_{\rm out}$ can change its sign if $GM^2 > e^2$. From the equation for turning points it follows that both roots are negative, i e. $0 \leq \rho < \infty$. This result is a rather evident because in the considered case both gravitation and Coulomb force are attractive. See Figs. 15 and 16.

The Case $\sigma_{\rm out}(\infty) = -1$ with Turning Point

In this case $\Delta m < 0$, $(\Delta m)^2 > M^2$, $\sigma_{\rm in}(\infty) = -1$, $\sigma_{\rm in}(0) = +1$. If the shell is falling into the T–region in the Schwarzschild metric ("out"), then the infall to the central singularity is inevitable. This means that the point of changing sign $\rho_\sigma(in)$ is always exists and is on the shell trajectory.

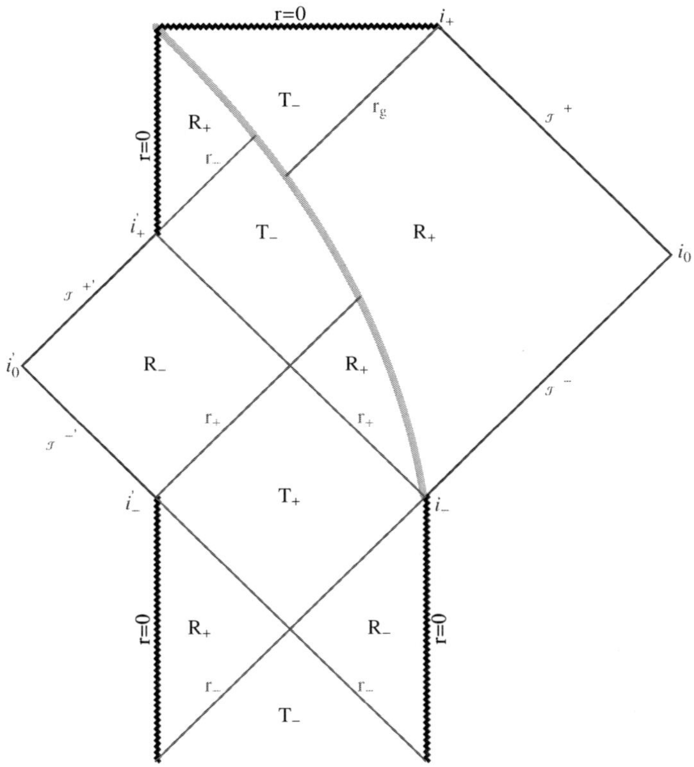

Figure 13. Infinite motion $\rho \to \infty$: the black hole case.

There is also a possibility that a turning point ρ_0 exists and is in the R–region of both internal and external metrics, and so $\rho_0 > r_g > r_+$. This means that a turning point is in the R_-–regions in the external ("out") and internal ("in") metrics. Respectively in this case the both σ_{in} and σ_{out} do not change signs. See Figs. 17.

The Case $\sigma_{out}(\infty) = -1$ without Turning Point

This corresponds to $B = 0$ in (59) or

$$\frac{\Delta m}{M} + GM(m + m_{in}) > 0 \tag{60}$$

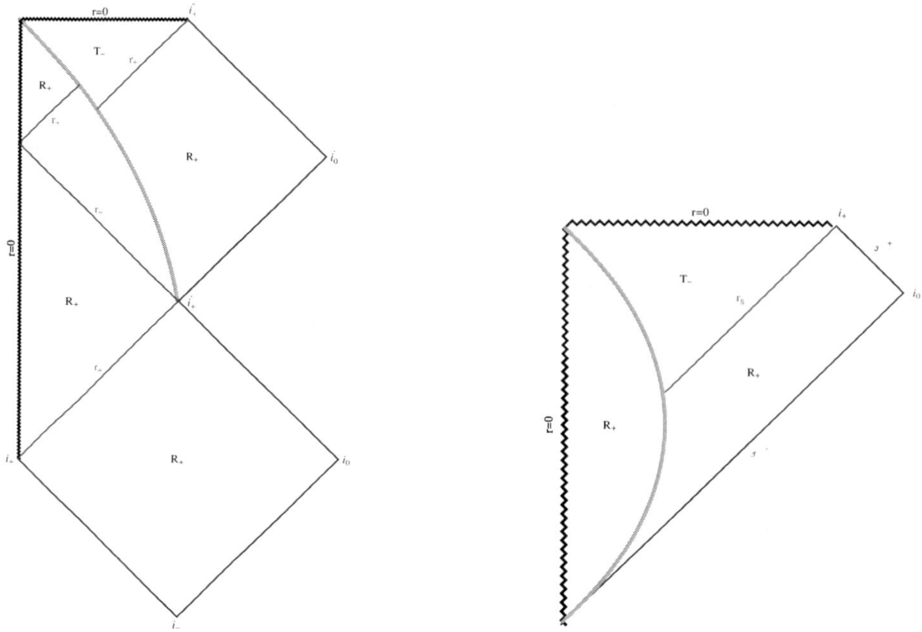

Figure 14. Infinite motion $\rho \to \infty$: cases of extreme black hole and naked singularity.

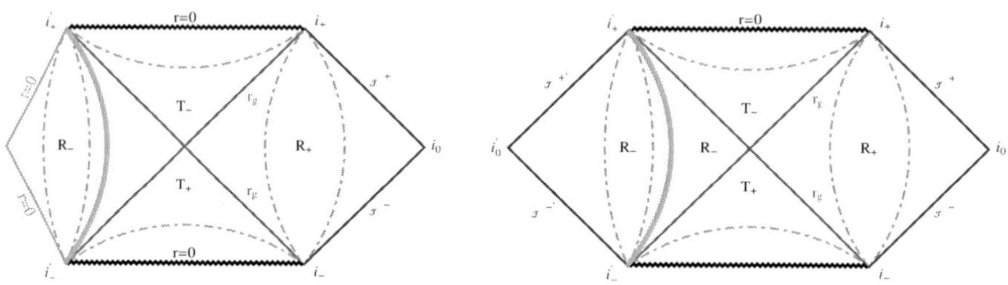

Figure 15. Infinite motion (Case 5.2.).

and

$$\frac{\Delta m}{M}(e^2 + GM^2) + 2GMm_{\text{in}} > 0. \tag{61}$$

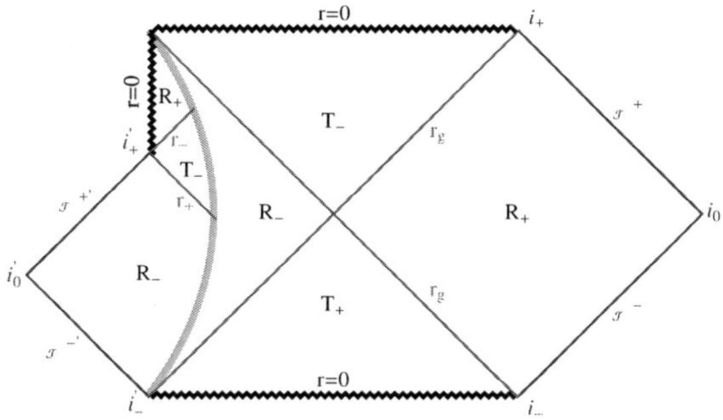

Figure 16. Infinite motion (Case 5.2.).

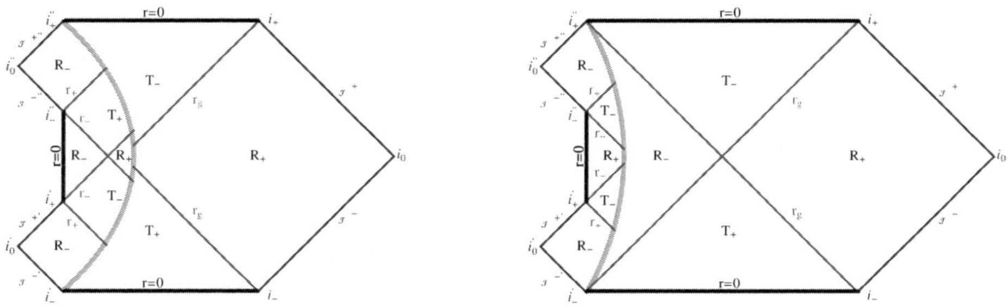

Figure 17. Infinite motion with turning point (Case 5.2.).

From these relations it follows

$$-\Delta m < 2m_{\text{in}}\left(1+\frac{e^2}{GM^2}\right)^{-1}. \tag{62}$$

In this case there are two negative roots for turning point:

$$\left[\frac{\Delta m}{M}(e^2+GM^2)+2GMm_{\text{in}}\right]^2 < (GM-e^2)^2\left[\frac{(\Delta m)^2}{M^2}-1\right]. \tag{63}$$

That is there are no turning point. The last inequality may be expressed as

$$\frac{(\Delta m)^2}{M^2}+\frac{\Delta m}{M}\frac{m_{\text{in}}}{M}\frac{e^2+GM^2}{e^2}+\frac{Gm_{\text{in}}^2}{M^2e^2}+\frac{(GM^2-e^2)^2}{4GM^2e^2} < . \tag{64}$$

$e^2 < Gm_{\text{in}}^2$. See Fig. 18.

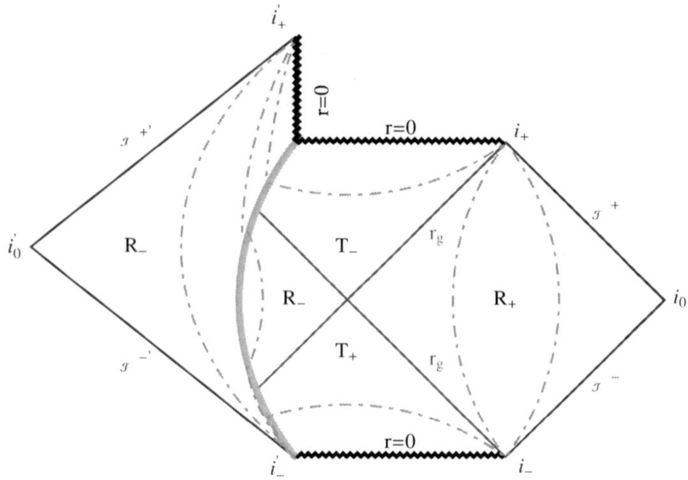

Figure 18. Case 5.2. with no turning point.

The Case $\sigma_{\text{out}}(\infty) = -1$ with Turning Point

Now we have

$$\frac{\Delta m}{M} + GM(m + m_{\text{in}}) < 0 \tag{65}$$

and

$$\frac{\Delta m}{M}(e^2 + GM^2) + 2GMm_{\text{in}} < 0. \tag{66}$$

From these relations it follows

$$-\Delta m > 2m_{\text{in}} \left(1 + \frac{e^2}{GM^2}\right)^{-1}. \tag{67}$$

There are two roots for turning point

$$\left[\frac{\Delta m}{M}(e^2 + GM^2) + 2GMm_{\text{in}}\right]^2 > (GM - e^2)^2 \left[\frac{(\Delta m)^2}{M^2} - 1\right]. \tag{68}$$

The roots with

$$\Delta m - \frac{GM^2 - e^2}{2\rho_0} \gtrless 0. \tag{69}$$

correspond respectively to $\sigma_{\text{in}}(\rho_0) = \pm 1$.

5.3. Infinite Motion Starting in R_-–region

It must be mentioned also the specific case of infinite motion starting in the R_-–region, where $\sigma_{\text{out}}(\infty) = -1$, with a turning point in the $R_+(in)$–region, where $\sigma_{\text{in}}(\rho_0) = +1$. This type of motion is realized if $GM^2 < e^2 < Gm_{\text{in}}^2$ and so $M < m_{\text{in}}$.

5.4. Finite Motion with $\rho < \infty$

$|\Delta m/M| < 1$.

The Case $\sigma_{\text{out}}(\rho_0) = +1$

In this case $\Delta m > 0$.

If $GM^2 > e^2$, then $\sigma_{\text{out}}(0) = -1$, From existence of ρ_σ it follows $\Delta m > 0$.

If $GM^2 < e^2$, then $\sigma_{\text{out}}(0) = +1$, ρ_σ does not exist and $\Delta m > 0$.

$$\Delta m - \frac{GM^2 - e^2}{2\rho_0} > 0. \tag{70}$$

$\sigma_{\text{in}} = +1$. See Fig. 19.

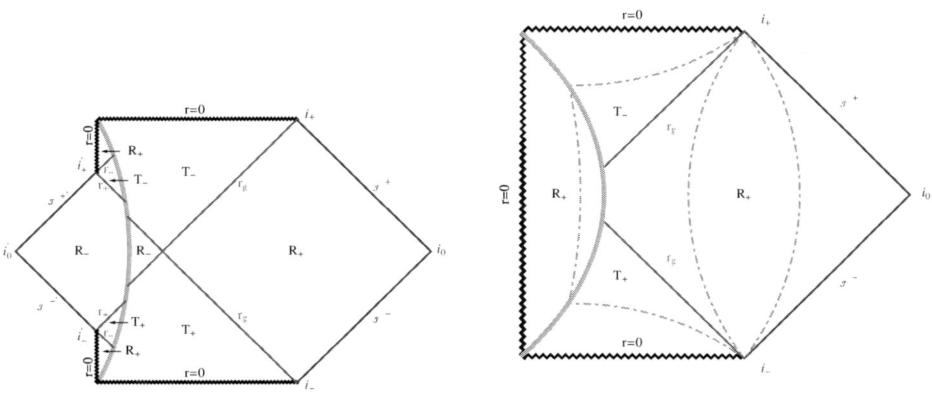

Figure 19. The Case 5.4. with a finite motion and $\Delta m > 0$.

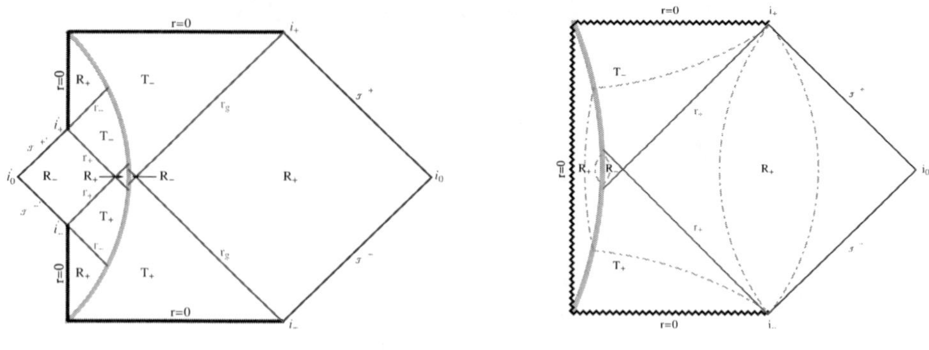

Figure 20. The Case 5.4. with a finite motion and $\Delta m > 0$.

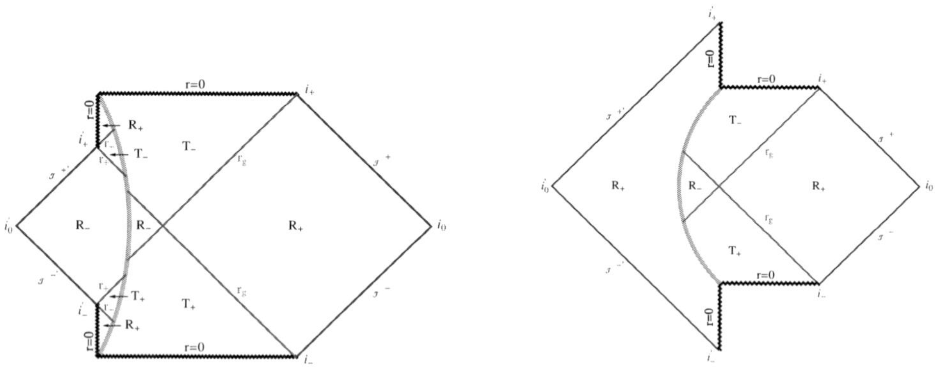

Figure 21. The Case of a naked singularity and $\Delta m > 0$.

The Case $\sigma_{\text{out}}(\rho_0) = -1$

$$\Delta m - \frac{GM^2 - e^2}{2\rho_0} < 0. \tag{71}$$

The case $\sigma_{\text{out}}(0) = -1$ corresponds to $GM^2 > e^2$. There are possible the both cases $\Delta m \gtrless 0$. In this case $\rho_\sigma > \rho_0$ and there is one root.

The case $\sigma_{\text{out}}(0) = +1$ corresponds to $GM^2 < e^2$ and so $\Delta m < 0$. See Fig. 20.

Chapter 6

Neutralizing Shell — Capacitor

We describe here the possible types of the spherically symmetric global geometries for the moving shell with the electric charge, which is equal and opposite to the corresponding charge of the internal Reissner–Nordström metric with mass $m_{\rm in}$ and electric charge e. This shell has a "naked" mass M and charge $-e$, which is neutralizing the charge of the internal source. As a result, the external metric is the Schwarzschild metric with the mass $m_{\rm tot} = m$.

The complete space-time consists of three parts. The first part — the internal (in) one is a piece of the Reissner–Nordström metric, defined by two parameters, the mass $m_{\rm in} > 0$ and charge e. The source of this metric may be the charged shell, described in the Part I. Here for simplicity we suppose that the sources of mass and charge are confined in the central singularity at $r = 0$.

The second part — the external (out) one is a piece of the Schwarzschild metric with the mass $m_{\rm tot} > 0$. This two parts are separated off each other by the third part — the thin shell with the a "naked" mass M and a compensating charge $-e$. The Carter–Penrose diagram for the total Schwarzschild space-time is shown in Fig. 2. The corresponding conformal diagrams for the Reissner–Nordström metric (see Figs. 3, 4 and 5) depends on the relation of parameters $e^2 \gtreqless Gm_{\rm in}^2$. The horizon radii are $r_\pm = Gm_{\rm in} \pm \sqrt{G^2m_{\rm in}^2 - Ge^2}$. In the case of the extreme black hole $r_- = r_+ = Gm_{\rm in}$. In the Carter–Penrose diagrams, the part of the Reissner–Nordström space-time manyfold would be at left from the shell (in), and the corresponding part of the Schwarzschild manyfold would be at right from the shell (out).

6.1. Equations

The corresponding Israel equation for the dynamics of the neutralizing shell

$$\sigma_{\rm in}\sqrt{\dot\rho^2+1-\frac{2Gm_{\rm in}}{\rho}+\frac{Ge^2}{\rho^2}}-\sigma_{\rm out}\sqrt{\dot\rho^2+1-\frac{2Gm_{\rm out}}{\rho}}=\frac{GM}{\rho}. \tag{72}$$

where the total mass $m_{\rm tot}=m_{\rm out}=m$, $\rho(\tau)$ — a shell radius as a function of the proper time τ, $\sigma=+1$ — radii are growing in the direction of the external normal to the shell, $\sigma=-1$ — the corresponding radii are diminishing. At the same time, $\sigma=+1$ at the R_+–regions, and, respectively, $\sigma=-1$ at R_-–regions. The sign of σ may be changed only in the $T_\pm$–regions, when a corresponding subradical expression in (72) is equal to zero. In this way $\sigma_{\rm out}$ is changing sign at the point $\rho_{\sigma_{\rm out}}$, for which are satisfied the following equations:

$$\dot\rho^2+1=\frac{2Gm_{\rm out}}{\rho_{\sigma_{\rm out}}},\quad \sigma_{\rm in}\sqrt{\frac{2G\Delta m}{\rho_{\sigma_{\rm out}}}+\frac{Ge^2}{\rho^2_{\sigma_{\rm out}}}}=\frac{GM}{\rho_{\sigma_{\rm out}}}, \tag{73}$$

where

$$\Delta m=m_{\rm out}-m_{\rm in}. \tag{74}$$

Certainly, this point may be absent on the specific shell trajectory. We see, that $\sigma_{\rm in}(\rho_{\sigma_{\rm out}})=1$ and, furthermore,

$$\Delta m = \frac{GM^2-e^2}{2\rho_{\sigma_{\rm out}}}>-\frac{e^2}{2\rho_{\sigma_{\rm out}}}, \tag{75}$$

$$\rho_{\sigma_{\rm out}} = \frac{GM^2-e^2}{2\Delta m},\quad ({\rm sign}[\Delta m]={\rm sign}[GM^2-e^2]). \tag{76}$$

At the same time, $\sigma_{\rm in}$ changes the sign at the point $\rho_{\sigma_{\rm in}}$, which is a solution of equations

$$\dot\rho^2+1=\frac{2Gm_{\rm in}}{\rho_{\sigma_{\rm in}}}-\frac{Ge^2}{\rho^2_{\sigma_{\rm in}}},\quad -\sigma_{\rm out}\sqrt{-\frac{2G\Delta m}{\rho_{\sigma_{\rm in}}}-\frac{Ge^2}{\rho^2_{\sigma_{\rm in}}}}=\frac{GM}{\rho_{\sigma_{\rm in}}}. \tag{77}$$

We see, that $\sigma_{\rm out}(\rho_{\sigma_{\rm in}})=-1$ and, furthermore,

$$\Delta m = -\frac{GM^2+e^2}{2\rho_{\sigma_{\rm in}}}<-\frac{e^2}{2\rho_{\sigma_{\rm in}}}<0, \tag{78}$$

$$\rho_{\sigma_{\rm in}} = \frac{GM^2+e^2}{2(-\Delta m)}. \tag{79}$$

At $\Delta m < 0$

$$\rho_{\sigma_{out}} = \frac{e^2 - GM^2}{2(-\Delta m)} \tag{80}$$

and, therefore, $\rho_{\sigma_{out}} < \rho_{\sigma_{in}}$. From the condition of "non exoticism" it follws, that at $\sigma_{in} = -1$ it is necessary must be $\sigma_{out} = -1$.

For investigation of the shell dynamics, it is needed to square the Israel equation (72):

$$\Delta m = M\sigma_{out}\sqrt{\dot{\rho}^2 + 1 - \frac{2Gm_{out}}{\rho}} + \frac{GM^2 - e^2}{2\rho}. \tag{81}$$

From here it follows, that

$$\sigma_{out} = \text{sign}\left[\Delta m - \frac{GM^2 - e^2}{2\rho}\right]. \tag{82}$$

With a help of (72) and (81) we find

$$\Delta m = M\sigma_{in}\sqrt{\dot{\rho}^2 + 1 - \frac{2Gm_{in}}{\rho} + \frac{Ge^2}{\rho^2}} - \frac{GM^2 + e^2}{2\rho}. \tag{83}$$

For physical interpretation of this equation it is useful to introduce the "running" mass $m_{run}(\rho)$ as an effective total mass (energy) inside the sphere of radius ρ. In our case this is a total energy with the deduction of energy, distributed beyond the sphere of radius ρ. For the inside metric this is

$$m_{in,run}(\rho) = m_{in} - \frac{e^2}{2\rho}, \tag{84}$$

i. e., the total mass at infinity (without the shell) with the deduction of electrostatic energy beyond the sphere. For the external metric this is

$$m_{out,run}(\rho) = m_{out}, \tag{85}$$

since outside the shell the electric field is absent. The difference of these running masses

$$\Delta m_{run} = m_{out,run} - m_{in,run} = \Delta m + \frac{e^2}{2\rho} \tag{86}$$

is a running mass of the shell. By substituting from here Δm to the squared equation, we have

$$\Delta m_{\rm run} = \sigma_{\rm in} M \sqrt{\dot\rho^2 + 1 - \frac{2Gm_{\rm run}}{\rho}} - \frac{GM^2}{2\rho}. \quad (87)$$

This equation in this form is viewed similar to the self-gravitating neutral shell with the only difference, that now $\Delta m_{\rm run}$ is already non constant. This is because the work of the Coulomb forces is not taken into account. The internal mass is changing due to the changing of the electrostatic energy, which influences the attraction inside the shell. Equation (87) at $\sigma_{\rm in} = +1$ has the sense of the energy conservation law. The term with the square root is an effective kinetic energy with addition of potential energy inside. The second term is an negative energy of the shell self-action. The kinetic energy formally changes sign if $\sigma_{\rm in} = -1$.

Up to now, we tacitly suppose the integration over the radius for defining the mass (energy) beyond the definite radius. The corresponding bare mass by definition is defined by integration along the direction of external normal of the shell, i. e. from left to right (subject to agreement) on the Carter–Penrose diagram. However, at $\sigma_{\rm in} = -1$ these directions are opposite. Therefore, though M is always positive, Δm may have any sign. Additionally, in the case $\sigma_{\rm in} = -1$ the center of the sphere (which is sometimes only imaginary) is placed beyond the sphere or at least not the inside. For this reason it is requested to change the sins in the definitions of as $m_{\rm run}$, and $\Delta m_{\rm run}$. In result, the shell is gravitationally attracted from inside to the outside (from left to right on the conformal diagram), however, as a matter of fact the shell is gravitationally attracted toward the (possible) center from the point of the internal observer. This is a quite realistic because the source is outside the shell. It must be taken into account for a qualitative understanding and physical interpretation of the described global geometries.

It is possible also the different interpretation. From the initial Israel equation it follows, that it is remained the same under the simultaneous changing of signs of the both σ with the additional changing the sign of the bare mass $M \to -M$. This transformation is equivalent to the exchange between (in) and (out), when sign M is changed automatically due to the changing of radial direction of integration.

The Israel equation is the Einstein constraint equation, integrated along the normal to the shell the Gauss coordinate. For fixed parameters of the internal metric m_{in} and e, and the shell parameters M_{in} and $-e$, the corresponding solution is defined by the initial conditions $\rho=\rho(0)$ and $\dot\rho=\dot\rho(0)$, $\sigma_{\text{in},0}=\sigma_{\text{in}}(\rho(0))$, $\sigma_{\text{out},0}=\sigma_{\text{out}}(\rho(0))$. At the same time, a total mass of the system, m_{out}, is calculated from the constraint equation. For infinite motion it is natural to define $\rho(0)=\infty$. With this determination $\sigma_{\text{in}}(\infty)$, $\sigma_{\text{out}}(\infty)$ and m_{out} define the shell velocity at infinity. For finite motion it useful to use the turning point ρ_0, which is fixing the other initial parameters: $\dot\rho_0=0$, $\sigma_{\text{in},0}=\sigma_{\text{in}}(\rho_0)$, $\sigma_{\text{out},0}=\sigma_{\text{out}}(\rho_0)$ and m_{out}. By putting $\rho_0=\infty$ in the squared equations, we have

$$\sigma_{\text{in}}(\infty)=\sigma_{\text{out}}(\infty)=\text{sign}[\Delta m]. \tag{88}$$

It is clear, that the cases $\Delta m>0$ and $\Delta m<0$ must be considered separately. For the qualitative description of the dynamical shell trajectory we need to know $\sigma(0)$, i. e., the value $\sigma(\rho)$ at $\rho=0$. We have

$$\sigma_{\text{out}}(0)=\text{sign}\left[e^2-GM^2\right],\quad \sigma_{\text{in}}(0)=\text{sign}\left[e^2+GM^2\right], \tag{89}$$

i. e., $\rho_{\sigma_{\text{out}}}$ at $\Delta m>0$ exists, if only $e^2>GM^2$, and, respectively, at $\Delta m<0$, if only $e^2<GM^2$. Accordingly, $\rho_{\sigma_{\text{in}}}$ exists only at $\Delta m<0$.

The infinite motion is realized at $(\Delta m/M)^2>1$, while the finite one at $(\Delta m/M)^2<1$. The turning points are defined from the twice squared Israel equation (72):

$$\frac{(GM^2-e^2)^2}{4\rho_0^2M^2}+\frac{1}{\rho_0 M}\left[\frac{\Delta m}{M}(e^2+GM^2)+2GMm_{\text{in}}\right]+\left(\frac{\Delta m}{M}\right)^2-1=0. \tag{90}$$

The roots of this equation are

$$\frac{1}{2\rho_0 M}=\frac{-B\pm\sqrt{D}}{(GM^2-e^2)^2}, \tag{91}$$

where

$$B = \frac{\Delta m}{M}(e^2-GM^2)+2GMm_{\text{out}}=\frac{\Delta m}{M}(e^2+GM^2)+2GMm_{\text{in}}, \tag{92}$$

$$D = B^2-(GM^2-e^2)^2\left[\left(\frac{\Delta m}{M}\right)^2-1\right]. \tag{93}$$

6.2. Neutralizing Shell at $\Delta m > 0$

We consider all possible types of shell trajectories at $\Delta m > 0$. In this case always $B > 0$, and roots of the quadratic equation are complex or negative at the infinite motion. Therefore, the turning points are absent. Under the finite motion — one of the roots is positive and other is negative, i. e., there is only turning point.

$\Delta m > 0$: Infinite Motion

In the infinite motion at $\Delta m > 0$ from the relation $\Delta m > M > 0$ follows that $M < m_{\rm out}$. Additionally, we have $\sigma_{\rm in} = +1$ — everywhere, $\sigma_{\rm out}(\infty) = +1$, $\sigma_{\rm out}(0) = {\rm sign}[e^2 - GM^2]$ and equation for the point, where $\sigma_{\rm out}$ changes the sign:

$$\rho_{\sigma_{\rm out}} = \frac{GM^2 - e^2}{2\Delta m}. \tag{94}$$

The point $\rho_{\sigma_{\rm out}}$ is absent, if $e^2 > GM^2$ (the self-repulsive shell). It corresponds to $\sigma(0) = +1$. On the contrary, for self-attractive shell, i. e., at $e^2 < GM^2$, the trajectory is inevitably has the point for changing the sign of σ, and $\sigma_{\rm out}(0) = -1$. Therefore, the self-repulsive shell is collapsing without obstruction.

In the second case it is possible the restriction due to the inequality $\rho_{\sigma_{\rm out}} < 2Gm_{\rm out}$. Let us verify this inequality:

$$\rho_{\sigma_{\rm out}} = \frac{GM^2 - e^2}{2\Delta m} < 2Gm_{\rm out} \tag{95}$$

$$GM^2 < G\Delta m^2 < 4G\Delta m^2 + 4G\Delta m\, m_{\rm in} + e^2. \tag{96}$$

It is clear that inequality is held automatically. The corresponding Carter–Penrose diagrams are shown in Figs. 22–23.

$\Delta m > 0$: Infinite Motion

Now consider the infinite motion at $\Delta m > 0$. It is a more complicated case, because of the possible existence of the turning point in $R_{\pm}$–regions, as in the internal and in the external metrics. Besides of this, σ may have the different signs at the point $r = 0$. In fact, we have

$$0 < \Delta m < M, \quad 0 \le \rho \le \rho_0. \tag{97}$$

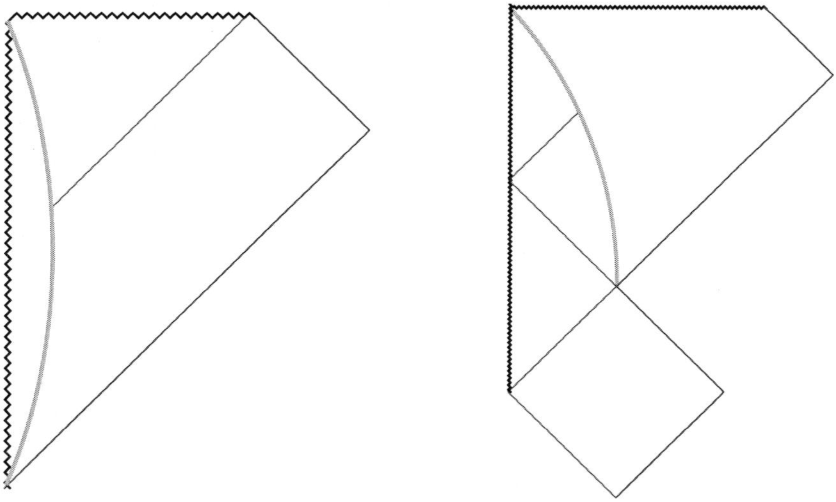

Figure 22. Infinite motion of the shell at $\Delta m > M$ and $e^2 > Gm_{in}^2$ (at the left panel) and $e^2 = Gm_{in}^2$ (at the right panel).

The turning point now is only one:

$$\frac{1}{2\rho_0 M} = \frac{\sqrt{D}-B}{(GM^2-e^2)^2}, \tag{98}$$

$$B = \frac{\Delta m}{M}(e^2-GM^2)+2GMm_{\text{out}} \tag{99}$$

$$= \frac{\Delta m}{M}(e^2+GM^2)+2GMm_{\text{in}}, \tag{100}$$

$$D = B^2+(GM^2-e^2)^2\left[1-\left(\frac{\Delta m}{M}\right)^2\right] \geq B^2. \tag{101}$$

In the limiting case $e^2 = GM^2$ we have

$$\frac{1}{2\rho_0 M} = \frac{1-\left(\frac{\Delta m}{M}\right)^2}{2B} = \frac{1-\left(\frac{\Delta m}{M}\right)^2}{4GMm_{\text{out}}}, \tag{102}$$

$$\rho_0 = \frac{2Gm_{\text{out}}}{1-\left(\frac{\Delta m}{M}\right)^2} > 2Gm_{\text{out}} = r_g. \tag{103}$$

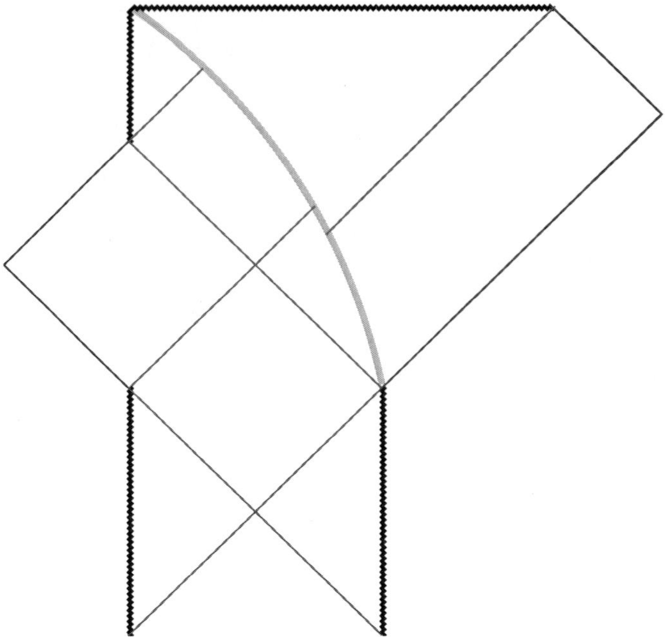

Figure 23. Infinite motion of the shell at $\Delta m > M$ and $e^2 < Gm_{in}^2$.

Let us verify this inequality in the general case, i. e., find any limitation to the parameters:

$$\frac{1}{2\rho_0 M} = \frac{\sqrt{D}-B}{(GM^2-e^2)^2} < \frac{1}{4GMm_{\rm out}} \quad \Rightarrow \quad (\sqrt{D}-2GMm_{\rm out})^2 \geq 0. \qquad (104)$$

Let us define condition, when turning point is placed directly at the horizon of the external metric:

$$\sqrt{D} = 2GMm_{\rm out} \quad \Rightarrow \quad (GM^2-e^2)(GM^2-e^2+4G\Delta m\, m_{\rm out}) = 0. \qquad (105)$$

The possibility $GM^2 - e^2 = 0$ is realized only, if additionally $\Delta m = 0$. However

$$\Delta m = \frac{e^2 - GM^2}{4Gm_{\rm out}}, \qquad (106)$$

Condition $\Delta m > 0$ is realized only in the case of the self-repulsive shell with $e^2 > GM^2$,

$$\Delta m < M \Rightarrow 0 < e^2 - GM^2 < 4GMm_{\rm out},\ GM^2 < e^2 < GM(M+4m_{\rm out}). \qquad (107)$$

Now define $\rho_{\sigma_{in}}$ and $\rho_{\sigma_{out}}$, which both determine the global geometry of the total space-time. From condition $\Delta m > 0$ it follows, that

$$\rho_{\sigma_{out}} = \frac{GM^2 - e^2}{2\Delta m} \tag{108}$$

exists only for the self-attractive shell with $GM^2 > e^2$. In this case $\sigma_{out}(0) = \text{sign}[e^2 - GM^2]$. In result, for the self-repulsive shell we have $\sigma_{out}(0) = +1$, and, therefore, the point ρ_σ is absent. This means, that turning point is inevitably placed in the R_+–region of the external metric. The conformal diagram for the Schwarzschild manifold at $e^2 > GM^2$ and $0 < \Delta m < M$ is shown in Fig. 24.

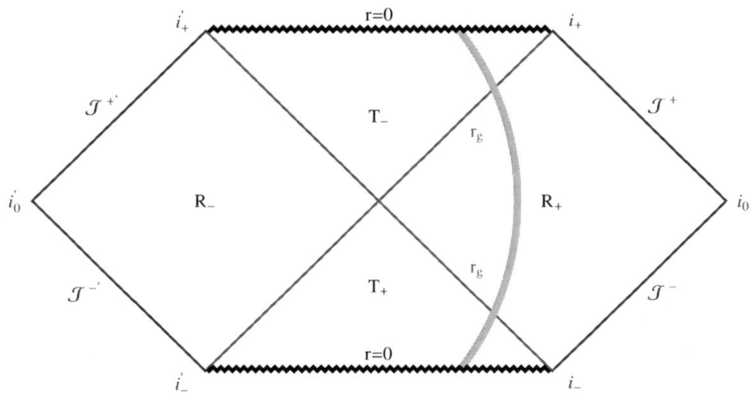

Figure 24. Finite motion of the shell in the external metric at $e^2 > GM^2$ and $0 < \Delta m < M$.

For the self-attractive shell $e^2 > GM^2$ and $\rho_\sigma(0) = -1$. The sign $\sigma(\rho_0)$ depends on the existence of the point $\rho_{\sigma_{out}}$ in the T–region of the external metric:

$$\rho_{\sigma_{out}} < 2Gm_{out} \Rightarrow \sigma(\rho_0) = +1, \tag{109}$$

$$\rho_{\sigma_{out}} > 2Gm_{out} \Rightarrow \sigma(\rho_0) = -1. \tag{110}$$

From equations (94) and (95) we have

$$\rho_{\sigma_{out}} = \frac{GM^2 - e^2}{2\Delta m} < 2GM_{out} \tag{111}$$

At $GM^2 > e^2$ this condition transforms to the inequality

$$\Delta m > \frac{m_{\rm in}}{2}\left(\sqrt{\frac{M^2}{m_{\rm in}^2} - \frac{e^2}{Gm_{\rm in}^2}} - 1\right). \tag{112}$$

Under this condition and, additionally, at $\Delta m < M$ we would have $\rho_\sigma(0) = +1$.

In the opposite case the point $\rho_{\sigma_{\rm out}}$ is placed in the R–region. In principle, the shell must come out into the R_-–region under realization of the conditions $\rho_0 < \rho_{\sigma_{\rm out}}$ and $\rho_{\sigma_{\rm out}} > 2Gm_{\rm out}$. These conditions are equivalent to the inequality

$$GM^2 > 4G\Delta m\, m_{\rm out} + e^2, \tag{113}$$

which by-turn corresponds to the condition $\rho_0 > r_g = 2Gm_{\rm out}$. In Fig. 25 are shown two conformal diagrams for the cases $GM^2 > e^2$ and

$$\Delta m \lessgtr \frac{m_{\rm in}}{2}\left(\sqrt{\frac{M^2}{m_{\rm in}^2} - \frac{e^2}{Gm_{\rm in}^2} + 1} - 1\right). \tag{114}$$

We see that this classification does not depend on the properties of the internal metric at $\Delta m > 0$, when $\sigma_{\rm in} = +1$ everywhere along the shell trajectory. The corresponding complete Carter–Penrose diagrams are shown in Figs. 26–28.

$\Delta m = 0$: Finite Motion

Before the description the more complicated and multivariate case $\Delta m < 0$, we consider the separately the specific interjacent case $\Delta m = 0$, and,conjointly, verify the previous formulas, as the limiting cases.

Forasmuch then, at the finite motion we have $\sigma_{\rm in} = +1$ and

$$\Delta m = 0 \quad \Rightarrow \quad \sigma_{\rm out} = \mathrm{sign}[e^2 - GM^2]. \tag{115}$$

At the same tine

$$B = 2GMm_{\rm out} = 2GMm_{\rm in} \tag{116}$$

$$D = 4GM^2 m_{\rm in}^2 + (GM^2 - e^2)^2, \tag{117}$$

$$\frac{1}{2\rho_0 M} = \frac{\sqrt{4GM^2 m_{\rm in}^2 + (GM^2 - e^2)^2} - 2GMm_{\rm out}}{(GM^2 - e^2)^2}, \tag{118}$$

$$GM^2 = e^2 \quad \Rightarrow \quad \rho_0 = 2GMm_{\rm out} = 2GMm_{\rm in}. \tag{119}$$

Consequently, the point ρ_0 is placed at the external horizon. The corresponding Carter–Penrose diagrams for the cases $e^2 \gtreqless GM^2$ are shown in Fig. 29.

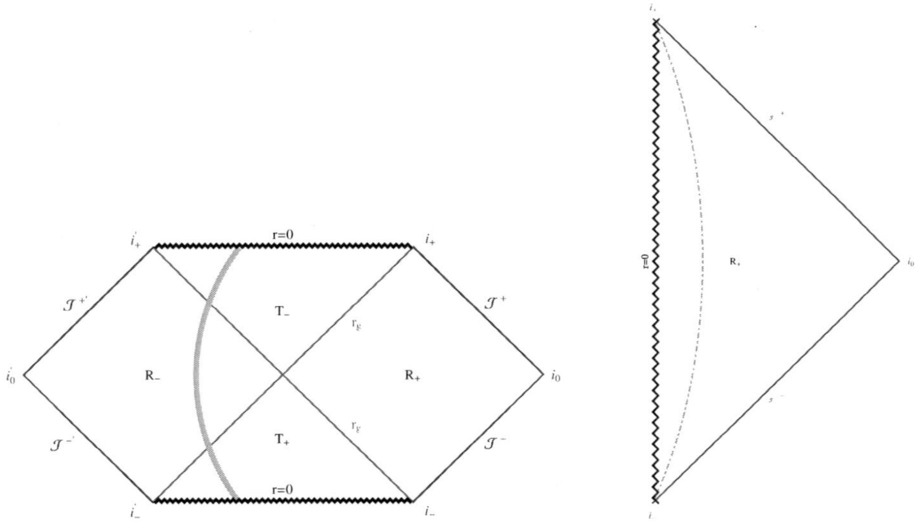

Figure 25. Finite motion of the shell in the external metric at $GM^2 > e^2$ and $\Delta m > (m_{\rm in}/2)[\sqrt{(M/m_{\rm in})^2 - (e^2/Gm_{\rm in}^2) + 1} - 1]$ at the down panel, and $\Delta m < (m_{\rm in}/2)[\sqrt{(M/m_{\rm in})^2 - (e^2/Gm_{\rm in}^2) + 1} - 1]$ at the upper panel.

Extreme Black Hole at $e^2 = GM^2$: Indifferent Shell

The one more interjacent case: $e^2 = GM^2$ — indifferent shell, which corresponds to conditions:

$$\sigma_{\rm out} = {\rm sign}[\Delta m], \quad \sigma_{\rm in} = {\rm sign}\left[\Delta m + \frac{GM^2}{\rho}\right], \tag{120}$$

$$B = D = 2GMm_{\rm out} > 0, \tag{121}$$

$$\rho_0 = \frac{2Gm_{\rm in}}{1 - \frac{\Delta m^2}{M^2}} > r_g. \tag{122}$$

Let us $\Delta m > 0$. In this case $\sigma_{\rm out} = \sigma_{\rm in} = +1$. The turning point is placed in the R_+–region of as external and the internal metrics. The form of the corresponding conformal diagrams is evident.

The case $\Delta m < 0$ is somewhat complicated because $\sigma_{\rm out} = -1$, and the sign

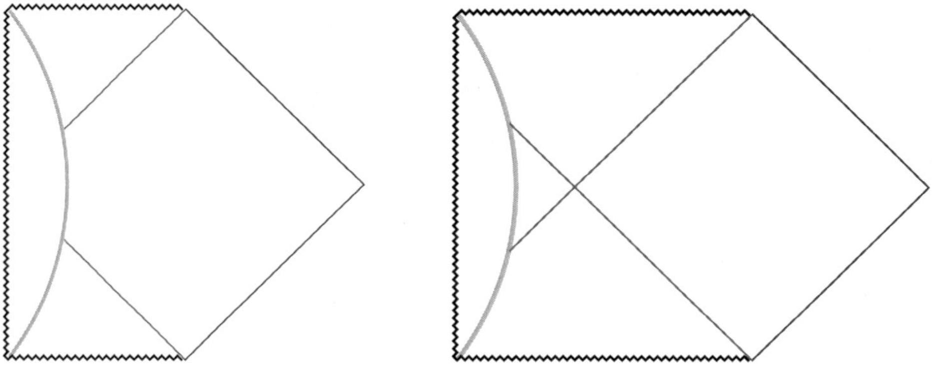

Figure 26. Finite motion: (the left panel) at $e^2 > GM^2$ and $0 < \Delta m/M < 1$ or at $e^2 > GM^2$ and $\Delta m > (m_{\rm in}/2)[\sqrt{(M/m_{\rm in})^2 - (e^2/Gm_{\rm in}^2) + 1} - 1]$, and (the right panel) at $e^2 < GM^2$, $0 < \Delta m/M < 1$ or at $e^2 > GM^2$ and $\Delta m < (m_{\rm in}/2)[\sqrt{(M/m_{\rm in})^2 - (e^2/Gm_{\rm in}^2) + 1} - 1]$.

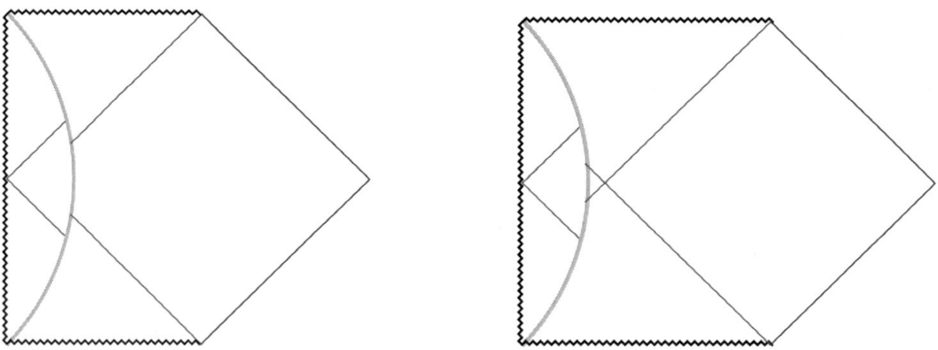

Figure 27. Finite motion of the shell: (the left panel) at $e^2 > GM^2$ and $0 < \Delta m/M < 1$ or at $e^2 < GM^2$ and $\Delta m > (M - m_{\rm in})/2$, and (the right panel) at $e^2 < GM^2$, $0 < \Delta m/M < 1$ or at $e^2 > GM^2$ and $\Delta m > (M - m_{\rm in})/2$.

$\sigma_{\rm in}$ may be changed. For the internal metric we have

$$\sigma_{\rm in}(\infty) = -1, \quad \sigma_{\rm in}(0) = +1, \tag{123}$$

$$\rho_{\sigma_{\rm in}} = -\frac{GM^2}{\Delta m}. \tag{124}$$

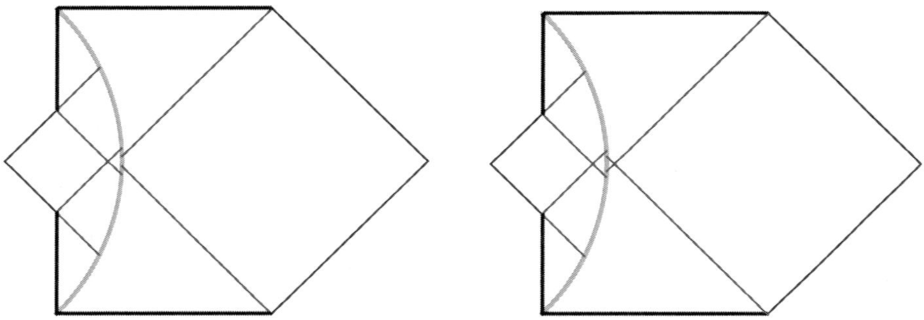

Figure 28. The case of finite motion of the shell at $\Delta m > 0$.

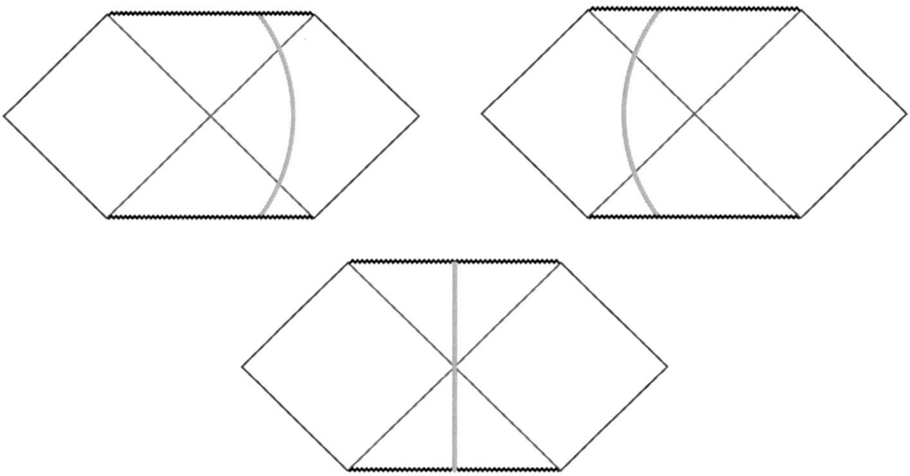

Figure 29. The case of finite motion at $\Delta m > 0$. The finite motion of the shell at $e^2 > GM^2$ (the left upper panel), $e^2 GM^2$ (the right upper panel) and $e^2 = GM^2$ (the central lower panel).

The problem is, first of all, for the infinite motion. On the one hand, the turning point is absent, and on the second hand, the sign σ must be changed. It is possible only in the T–region, which is absent, if $e^2 \geq Gm_{\rm in}^2$. We come to contradiction:

$$
\begin{aligned}
e^2 &= GM^2 \quad \Rightarrow \quad M \geq m_{\text{in}} && (125)\\
\Delta m &< 0 \quad \Rightarrow \quad m_{\text{out}} < m_{\text{in}}, \quad M \geq m_{\text{in}} > m_{\text{out}}, && (126)\\
\Delta m &< -M, \quad 0 < m_{\text{out}} < m_{\text{in}} - M \quad \Rightarrow \quad m_{\text{in}} > M. && (127)
\end{aligned}
$$

Consequently, the infinite motion at $\Delta m < 0$, $e^2 = GM^2 \geq Gm_{\text{in}}^2$ is forbidden by condition $m_{\text{out}} > 0$.

For the possibility of infinite motion at $e^2 = GM^2 < Gm_{\text{in}}^2$ it is required that the turning point $\rho_{\sigma_{\text{in}}}$ must be between the two horizons, $r_- < \rho_{\sigma_{\text{in}}} < r_+$. Condition $M > 0$ at $\sigma_{\text{in}} = \sigma_{\text{out}} = +1$ (i. e. in the R_-–region) demands that

$$
\Delta m < -\frac{e^2}{\rho} = -\frac{GM^2}{\rho}. \tag{128}
$$

At $|\Delta m < M|$ it is followed, that $\rho < GM$. This means that in this case the infinite motion is impossible.

Consider now the finite motion at $\Delta m < 0$ and $\Delta m/M > -1$. For the turning point we have

$$
\sigma_{\text{in}}(\rho_0) = \text{sign}\left[\Delta m + \frac{GM^2}{\rho_0}\right] = \text{sign}\left[\Delta m + \frac{M^2\left(1 - \frac{\Delta m^2}{M^2}\right)}{2m_{\text{out}}}\right]. \tag{129}
$$

$$
\sigma(\rho_0) = \text{sign}[M^2 - \Delta m^2 + 2m_{\text{out}}\Delta m]. \tag{130}
$$

Roots of equation

$$
M^2 - \Delta m^2 + 2m_{\text{out}}\Delta m = 0 \tag{131}
$$

are

$$
\left(\frac{\Delta m}{M}\right)_{\pm} = -\frac{m_{\text{in}}}{M} \pm \sqrt{\frac{m_{\text{in}}^2}{M^2} - 1}. \tag{132}
$$

$$
e^2 = GM^2 < GM^2 m_{\text{in}} \quad \Rightarrow \quad M < m_{\text{in}}. \tag{133}
$$

It turns out, that $\sigma_{\text{in}}(\rho_0) = +1$, if

$$
0 > \frac{\Delta m}{M} > \sqrt{\frac{m_{\text{in}}^2}{M^2} - 1} - \frac{m_{\text{in}}}{M}, \tag{134}
$$

and $\sigma_{\text{in}}(\rho_0) = -1$, if

$$-1 < \frac{\Delta m}{M} < \sqrt{\frac{m_{\text{in}}^2}{M^2} - 1} - \frac{m_{\text{in}}}{M}. \tag{135}$$

Now it is requested to examine where are placed $\rho_{\sigma_{\text{in}}}$. $\rho_{\sigma_{\text{in}}} < \rho_0$:

$$-\frac{GM^2}{\Delta m} < \frac{2Gm_{\text{out}}2}{1 - \frac{\Delta m^2}{M^2}} \quad \Rightarrow \quad M^2 + \Delta m^2 - 2\Delta m\, m_{\text{in}} < 0. \tag{136}$$

This inequality is held for $\sigma_{\text{in}}(\rho_0) = -1$. Consequently, the sign σ_{in} is changing on the trajectory. In the second case the sign σ_{in} is not changed on the trajectory. If $e^2 = GM^2 > Gm_{\text{in}}^2$, then $M > m_{\text{in}}$, and in this case $\sigma_{\text{in}}(\rho_0) = +1$ without any limitations.

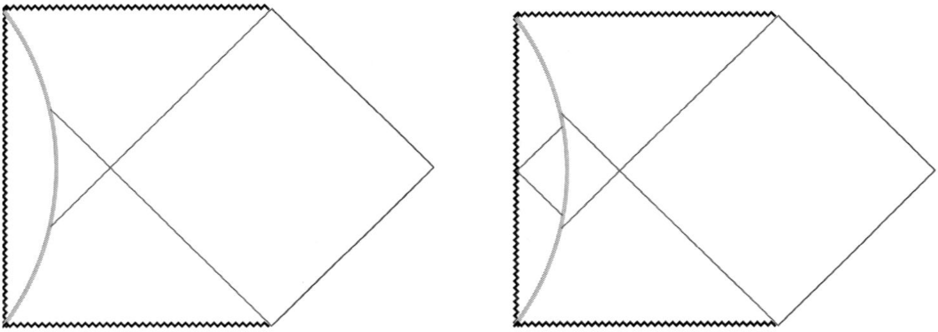

Figure 30. Finite motion of the shell at $e^2 = GM^2$. The left panel: at $e^2 = GM^2 > Gm_{\text{in}}^2$. The right panel: at $e^2 = GM^2 = Gm_{\text{in}}^2$.

The conformal diagrams for finite motion at $e^2 = GM^2$ and $\Delta m < 0$ are shown in Figs. 30–31 for the cases $e^2 = GM^2 > GM^2 m_{\text{in}}$, $e^2 = GM^2 = GM^2 m_{\text{in}}$ and $e^2 = GM^2 < GM^2 m_{\text{in}}$ at

$$0 > \frac{\Delta m}{M} > \sqrt{\frac{m_{\text{in}}^2}{M^2} - 1} - \frac{m_{\text{in}}}{M}. \tag{137}$$

and

$$-1 < \frac{\Delta m}{M} < \sqrt{\frac{m_{\text{in}}^2}{M^2} - 1} - \frac{m_{\text{in}}}{M}. \tag{138}$$

The infinite motion in this case is impossible.

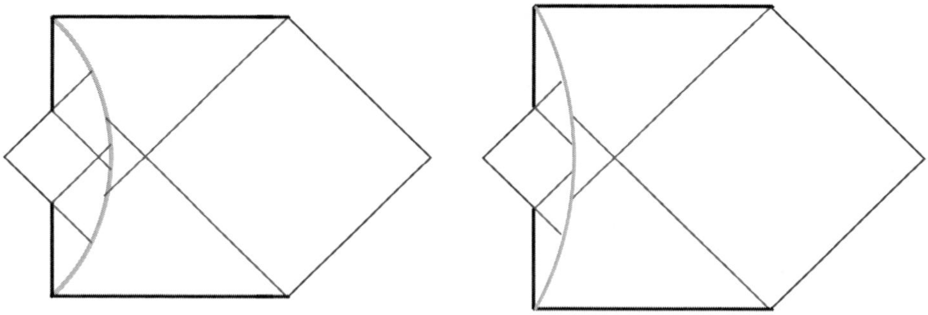

Figure 31. Finite motion of the shell at $e^2 = GM^2 < Gm_{\rm in}^2$. The left panel: at $0 > \Delta m/M > \sqrt{(m_{\rm in}^2/M)^2 - 1} - (m_{\rm in}/M)$. The right panel: at $-1 < \Delta m/M < \sqrt{(m_{\rm in}^2/M)^2 - 1} - (m_{\rm in}/M)$.

6.3. Neutralizing Shell at $\Delta m < 0$

We proceed now to the description of the possible alternative cases for the motion of the neutralizing shell at $\Delta m < 0$. Mow the sign $\sigma_{\rm in}$ may be changed. For this reason the number of possible combinations of parameters is greatly increased. Now the internal game is going into the game and we need to consider the different types of the Reissner–Nordström metric: $e^2 > Gm_{\rm in}$ — naked singularity, $e^2 < Gm_{\rm in}^2$ — black hole, and $e^2 = Gm_{\rm in}^2$ — extreme black hole.

We start our analysis from the simplest case from the point of view of the global Reissner–Nordström metric with naked singularity.

Naked Singularity at $e^2 > Gm_{\rm in}^2$: Infinite Motion

As long as

$$\sigma_{\rm in} = \mathrm{sign}[\Delta m + \frac{GM^2}{\rho}], \tag{139}$$

then $\sigma_{\rm in}(0) = +1$ and $\sigma_{\rm in}(\infty) = -1$. Therefore, at the infinite motion, starting at $\rho = \infty$, the trajectory can not reach the radial point $\rho = 0$. In the case of the naked singularity the T–region is absent, and so it is absent the point, where $\sigma_{\rm in}$ may change the sign. For this reason it must inevitably the turning point in the R_-–region with a subsequent motion again to infinity. For this reason, the finite motion begins at $\rho = 0$ in the R_+–region and must have the turning point at the same R_+–region (all from the point of you of the internal metric). In this case the conformal diagram for the internal metric has the form, shown in at the left panel in Fig. 33. These qualitative conclusions must be still proved, i. e. it is requested to prove the real existence of these trajectories and define the necessary relations between the involved parameters: M, Δm and $m_{\rm out}$ at the fixed $m_{\rm in}$ and e.

At first consider relations for infinite motion of the shell:

$$\Delta m/M < -1 \quad \Rightarrow \quad m_{\rm in} > M, \tag{140}$$
$$e^2 > Gm^2_{\rm in} \quad \Rightarrow \quad e^2 > GM^2. \tag{141}$$

This means, that in this case the shell is self-repulsive due to existence of the turning point. Further, let us see the equation for the turning point (91). The both functions B and D may have, in principle, the different signs, since $\Delta m/M < -1$. We need to have $B < 0$ and $D > 0$. Only in this case there are two real roots and both of them are positive. We demonstrate that this is really the case. Starting from expression for B in equation (93):

$$B = \frac{\Delta m}{M}(e^2+GM^2)+2GMm_{\rm in}, \tag{142}$$
$$= \left(\frac{\Delta m}{M}+1\right)(e^2+GM^2)-(e^2-Gm^2_{\rm in})-(Gm^2_{\rm in}-M)^2. \tag{143}$$

It is evident, that now $B < 0$. Next, let us transform the expression (93) for D:

$$D = B^2-(GM^2-e^2)\left(\frac{\Delta m^2}{M^2}-1\right) \tag{144}$$
$$= Ge^2\left[2\Delta m+\frac{m_{\rm in}}{e^2}(e^2+GM^2)\right]^2+(GM^2-e^2)^2\left(1-\frac{Gm^2_{\rm in}}{e^2}\right) > 0, \tag{145}$$

since $e^2 > Gm^2_{\rm in}$. We obtain for the used combination of parameters that equation for turning point really has two positive roots (formally there are two turning point: $\rho = \rho_{0,-}$ and $\rho = \rho_{0,+} \geq \rho_{0,-}$). It is clear, starting from infinity, the shell at first reaches the turning point, corresponding to the bigger root $\rho = \rho_{0,+}$, and after that again is moving toward the radial infinity. We need to clarify where it is placed the the point of the sign changing $\sigma_{\rm in}$, i. e., is it identically valid $\rho_{\sigma_{\rm in}} < \rho_{0,+}$, or there is additional constraint to parameters? Point is that

$\rho_{\sigma_{\rm in}}$ always exists at $\Delta m < 0$, but the shell cannot reach this radius because the necessary for sign changing T–region is absent at ll in the case of the naked singularity. Thus,

$$\frac{1}{2\rho_{0,\pm}M} = \frac{-B \pm \sqrt{D}}{(e^2 - GM^2)^2}, \quad \rho_{\sigma_{\rm in}} = -\frac{e^2 + GM^2}{2\Delta m} \tag{146}$$

To find the point $\rho_{\sigma_{\rm in}}$ we consider the square trinomial $A(\rho)$, With roots, corresponding to the turning points:

$$A(\rho) = \frac{(e^2 - GM^2)^2}{4\rho^2 M^2} + \frac{2B}{2\rho M} + \frac{\Delta m^2}{M^2} - 1. \tag{147}$$

The corresponding graphs is shown in Fig. 32. The value of $A(\rho)$ at the point

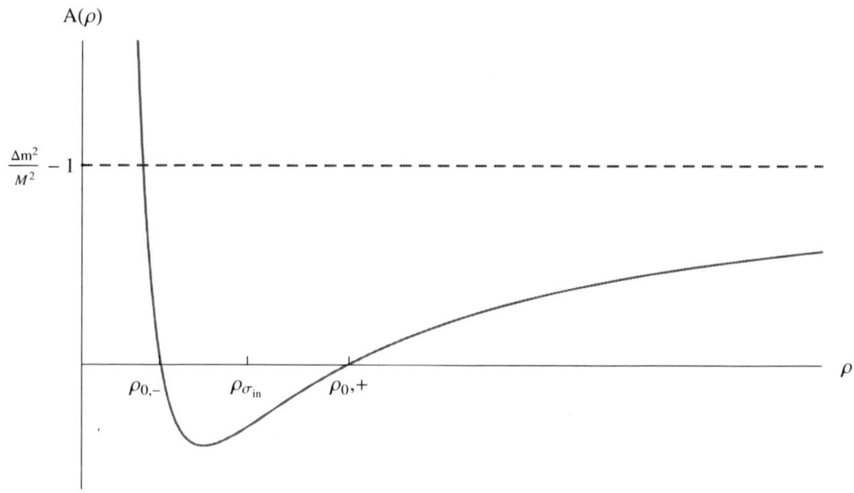

Figure 32. The square trinomial $A(\rho)$ from equation (147).

$\rho_{\sigma_{\rm in}}$ is

$$A(\rho_{\sigma_{\rm in}}) = -\frac{G{m_{\rm in}}^2}{e^2}\left(\frac{\Delta m}{M}\frac{e^2}{e^2 + GM^2} + 1\right)^2 - \left(1 - \frac{G{m_{\rm in}}^2}{e^2}\right) < 0. \tag{148}$$

It is remarkable that now $\rho_{\sigma_{\rm in}}$ is just placed between bigger and smaller turning points.

In that way we see, that infinite motion at $\Delta m/M < -1$ and $e^2 > Gm_{\rm in}^2$ is possible only for the self-repulsive shell. In this case the conformal diagram for the internal metric has the form, shown in at the left panel in Fig. 33. At the same time, the point for sign changing of $\sigma_{\rm out}$ exists, in which connection $\rho_{\sigma_{\rm out}} < \rho_{\sigma_{\rm in}}$. That is why the shell, moving from infinity does not reach the turning point $\rho_{\sigma_{\rm out}}$ and so $\sigma_{\rm out} = -1$. The corresponding complete conformal diagram is shown at he rigt panel in Fig. 33. Left form the shell at this diagram there is not an electro-vacuum space-time with the infinity, but someone source with a total mass $m_{\rm in}$ and electric charge e. For example, it may be the thin shell, again with the Schwarzschild metric at left side from the shell.

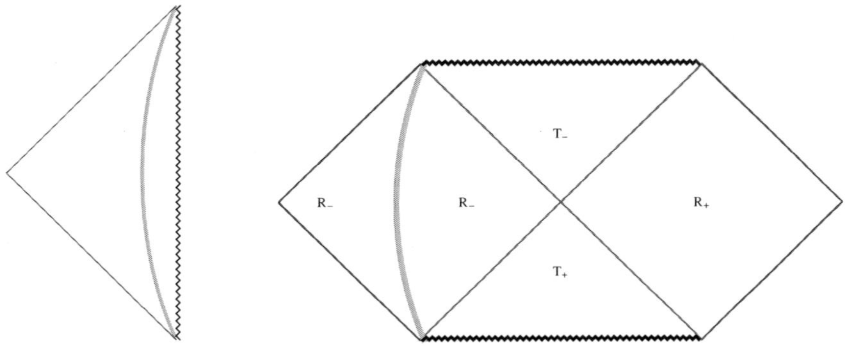

Figure 33. Infinite motion of the self-repulsive shell at $\Delta m/M < -1$ and $e^2 > Gm_{\rm in}^2$. The left panel: internal metric. The right panel: the complete metric.

A behavior of the parabolic trajectory at $\Delta m/M = -1$ is quite similar to the previously discussed the hyperbolic one. At this point we finish the description of the infinite motion.

Naked Singularity $e^2 > Gm_{\rm in}^2$: Finite Motion

Now we start a consideration of the finite motion at

$$-1 < \frac{\Delta m}{M} < 0, \quad e^2 > Gm_{\rm in}^2. \tag{149}$$

Since $\sigma_{\rm in}(0) = +1$, and T–region is absent, then $\sigma_{\rm in} = +1$ everywhere at the trajectory. The corresponding equation for the turning points has one positive and one negative root, i. e., there is only one turning point, which is physically acquitted.

The turning point ρ_0 in the equation (91) corresponds to the positive value of the discriminant D, but B may have any sign. Herewith there is point $\rho_{\sigma_{\rm in}}$ for sign changing in $\sigma_{\rm in}$, but we need to verify that $\rho_{\sigma_{\rm in}} > \rho_0$. As previously, $A(\rho_{\sigma_{\rm in}}) < 0$, and graph of the function $A(\rho)$ from (147) is shown in Fig. 34. Note, that now there is possible as in the case $e^2 > GM^2$ (self-repulsive shell) and $e^2 < GM^2$ (self-attractive shell).

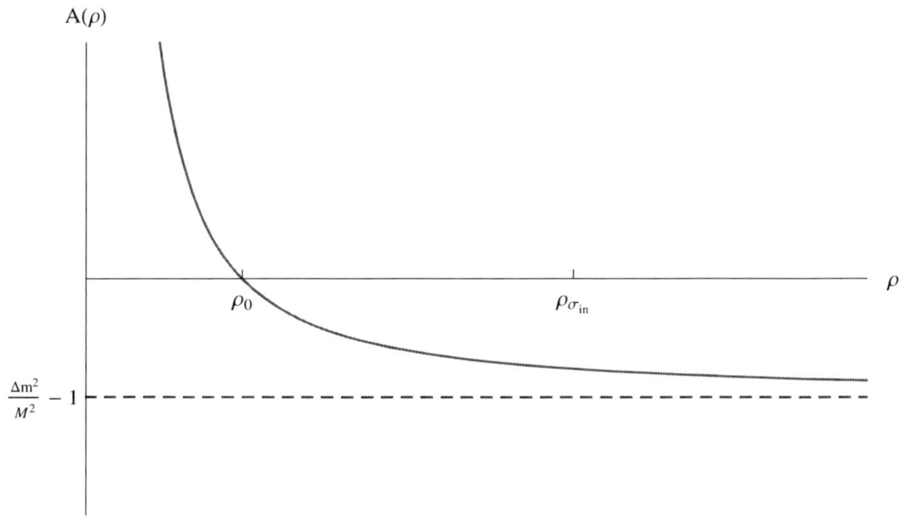

Figure 34. Graph of the function $A(\rho)$ from (147) for the case of the finite motion of the shell with one turning point.

We start from the case $e^2 < GM^2$, when $\rho_{\sigma_{\rm out}}$ does not exist and $\sigma_{\rm out}(0) = -1$. This mean, that the moving shell has turning point in the R_-–region of the external metric. The complete conformal diagram shown in Fig. 35.

If $e^2 > GM^2$, then $\rho_{\sigma_{\rm out}}$ exists because $\Delta m < 0$ and $\sigma_{\rm out}(0) = 1$. Now there are possible two cases:

(1) $\rho_{\sigma_{\rm out}}$ is placed in the T_-–region. The shell is starting at $\rho = 0$, then σ changes its sign and the shell is coming to the R_-–region, where there is turning point for the trajectory;

(2) $\rho_{\sigma_{\rm out}}$ is placed outside of the event horizon, i. e., $\rho_{\sigma_{\rm out}} > 2Gm_{{\rm out}=r_g}$. In

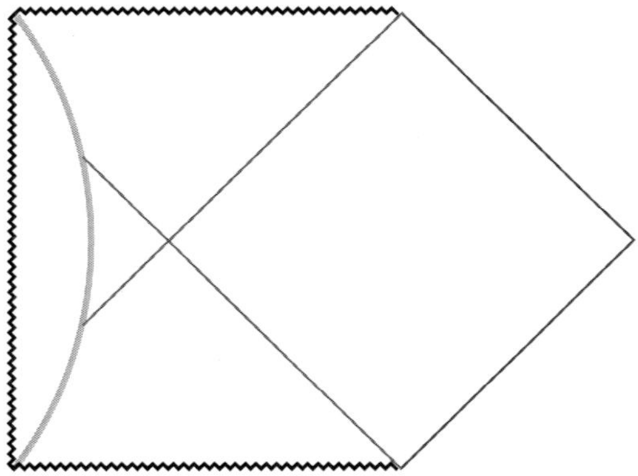

Figure 35. Conformal diagram for the case $e^2 < GM^2$, when $\rho_{\sigma_{\rm out}}$ does not exist and $\sigma_{\rm out}(0) = -1$.

this case the turning point is placed in the R_+–region. Besides, it is held true $\rho_{\sigma_{\rm out}} > \rho_0$.

Consider the inequality

$$\rho_{\sigma_{\rm out}} > r_g \quad \Rightarrow \quad -M < \Delta m < -\frac{e^2 - GM^2}{4Gm_{\rm out}}. \tag{150}$$

In this case the turning point is placed in the R_-–region. If it is satisfied

$$-\frac{e^2 - GM^2}{4Gm_{\rm out}} < \Delta m, \tag{151}$$

then the turning point is placed in the R_+–region.

Now we find the location of the turning point ρ_0. To do this we calculate

$$A(\rho_{\sigma_{\rm out}}) = 1 - \frac{4\Delta m\, m_{\rm out}}{e^2 - GM^2} < 0. \tag{152}$$

Since $A(\rho_{\sigma_{\rm out}}) < 0$, then $\rho_{\sigma_{\rm out}} > \rho_0$, and the shell is moving without changing the sign of σ. The finite motion is realized at $-M < \Delta m < 0$. This means that $m_{\rm in} - M < m_{\rm out} < m_{\rm in}$ and, consequently,

$$-M < \Delta m < -\frac{e^2 - GM^2}{4Gm_{\rm out}} \tag{153}$$

In this case there is a turning point in the R_-–region. At

$$0 > \Delta m > -\frac{e^2 - GM^2}{4Gm_{\rm out}} \tag{154}$$

The shell is is passing through the R_+–region of the external metric. At last, at

$$m_{\rm out} < \frac{e^2 - GM^2}{4GM}, \tag{155}$$

the shell trajectory always has the turning point in the R_+–region.

Extreme Black Hole at $e^2 = Gm_{\rm in}^2$: Infinite Motion

In the case of the extreme Reissner–Nordström black hole with $e^2 = Gm_{\rm in}^2$ for internal metric we have

$$B = \left(\frac{\Delta m}{M} + 1\right)(e^2 + GM^2) - (e - \sqrt{G}M)^2 \tag{156}$$

$$= \frac{\Delta m}{M}(e^2 + GM^2) + 2\sqrt{G}Me, \tag{157}$$

$$D = \left(2e\Delta m + e^2 + GM^2\right)^2 > 0. \tag{158}$$

For turning point we find

$$\frac{1}{2\rho_0(\pm)M} = \frac{-\frac{\Delta m}{M}(e^2 + GM^2) - 2\sqrt{G}Me \pm |2e\sqrt{G}\Delta m + e^2 + GM^2|}{(e^2 - GM^2)^2} \tag{159}$$

The difference of the case, considered previously for the naked singularity is the possibility for turning point to be placed as beyond and also inside of the double horizon of the internal metric $r_+ = r_- = Gm_{\rm in} = \sqrt{G}e$. The discriminant D now may be equal to zero at

$$\Delta m = -\frac{e^2 + GM^2}{2e\sqrt{G}}. \tag{160}$$

In this case we find that $\rho_0(+) = \rho_0(-) = r_\pm$, i. e., the double turning point is placed exactly at the double horizon. As before, in the case of infinite motion there are two positive roots, while at the finite motion one root is positive and the second one is negative.

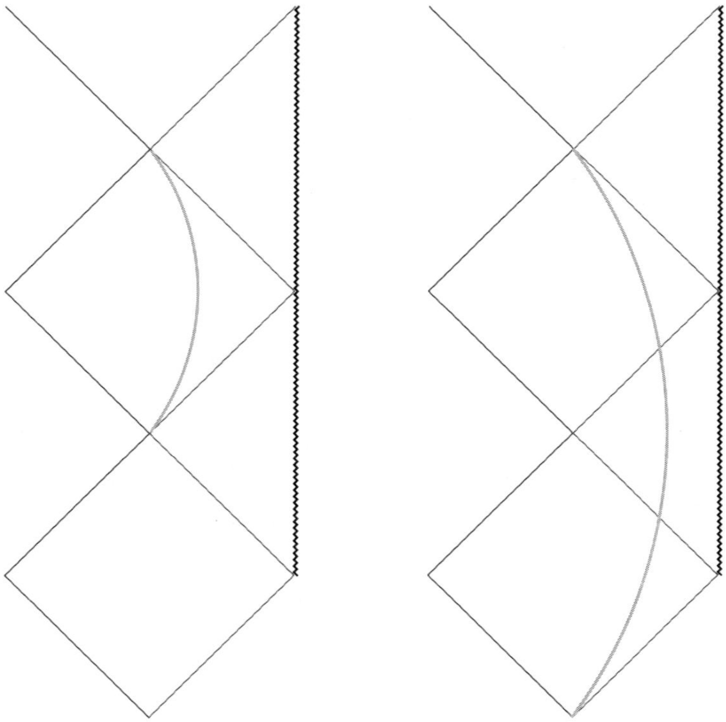

Figure 36. Internal metric with extreme black hole. Infinite motion with the turning point.

We start our analysis from the infinite motion at $\Delta m < -M$ when $\sigma_{\text{in}}(\infty) = \text{sign}(\Delta m) = -1$. The corresponding Carter–Penrose diagram for the internal metric is shown in Fig. 36 (at the left panel).

Let us initially

$$-\frac{e^2 + GM^2}{2e\sqrt{G}} < \Delta m < -M. \tag{161}$$

We verify now that these inequalities are compatible:

$$-\frac{e^2 + GM^2}{2e\sqrt{G}} < -M \quad \Rightarrow \quad -(e - \sqrt{G}M)^2 \leq 0. \tag{162}$$

We find for the turning point

$$\rho_0(\pm) = \frac{(e \pm \sqrt{G}M)^2}{2M\left(-\frac{\Delta m}{M} \pm 1\right)}. \tag{163}$$

Let us show that at

$$-\frac{e^2 + GM^2}{2e\sqrt{G}} < \Delta m \tag{164}$$

it is held the relation. Really, we have

$$\frac{(e - \sqrt{G}M)^2}{2M\left(-1 - \frac{\Delta m}{M}\right)} > \frac{(e + \sqrt{G}M)^2}{2M\left(1 - \frac{\Delta m}{M}\right)} \quad \Rightarrow \quad \Delta m > -\frac{e^2 + GM^2}{2\sqrt{G}e}, \tag{165}$$

as it must be. Similar to the case of the naked singularity, the point of the changing sign of σ is placed between two roots. This is easily verified by the direct calculations:

$$A(\rho_\sigma) < 0 \quad \Rightarrow \quad \rho_0(+) < \rho_\sigma < \rho_0(-). \tag{166}$$

For the point of the double horizon $r_\pm$ we find:

$$A(r_\pm) = \left(\frac{e^2 + GM^2}{2GMr_\pm} + \frac{\Delta m}{M}\right)^2. \tag{167}$$

By the direct calculations we verify that

$$r_\pm < \rho_0(+) \quad \Rightarrow \quad \Delta m > -\frac{e^2 + GM^2}{2\sqrt{G}e}, \tag{168}$$

i. e., at $\Delta m > -(e^2 + GM^2/(2\sqrt{G}e)$ the smaller turning point $\rho_0(+)$ is placed outside the double horizon. The shell, starting at infinity, is reaching the turning point, which is the larger root, and further is moving to infinity without meeting the smaller turning point. The corresponding diagrams is shown in Fig. 36 (at the left panel). At $\Delta m/M \to \infty$ this turning point tends to infinity, and at $\Delta m \to -(e^2 + GM^2/(2\sqrt{G}e)$ the turning point tends to the double horizon, where the both turning points are also doubling. At $\Delta m = -(e^2 + GM^2/(2\sqrt{G}e)$, as we already know, $\rho_0(+) = \rho_0(-) = \rho_\sigma = r_\pm$.

In the event $\Delta m < -(e^2 + GM^2/(2\sqrt{G}e)$, the turning points $\rho_0(+)$ and $\rho_0(-)$ switch their positions, i. e.,

$$\rho_0(\pm) = \frac{(e \mp \sqrt{G}M)^2}{2M\left(-\frac{\Delta m}{M} \mp 1\right)}, \tag{169}$$

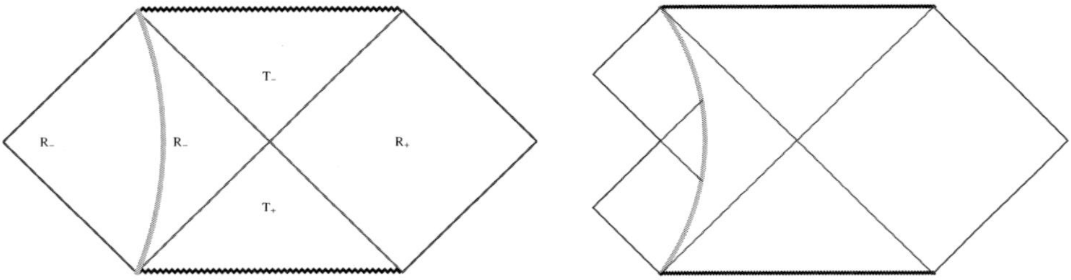

Figure 37. The case of extreme black hole. Infinite motion with the turning point: at $-(e^2+GM^2/(2\sqrt{G}e) < \Delta m < -M$ (at the left panel) and at $\Delta m < -(e^2+GM^2/(2\sqrt{G}e) < -M$ (at the right panel).

As before, $\rho_0(+) < \rho_\sigma < \rho_0(-)$, however, $r_\pm > \rho_0(-)$, and, consequently, the both turning points are placed inside the double horizon. Remind, that from $\Delta m < -M$ it follows that $M < m_{\text{in}} = e/\sqrt{G}$. The corresponding diagrams is shown in Fig. 36 (at the right panel).

For the external metric at $e^2 > GM^2$ there exists the point $\rho_{\sigma_{\text{out}}}$. It is evident, that $\rho_{\sigma_{\text{out}}} < \rho_{\sigma_\in} < \rho_0$, where ρ_0 — is the larger root, and, so $\sigma_{\text{out}}(\rho_0) = -1$. The corresponding conformal diagrams ar show in Fig. 37. At the left panel it is clear seen, that $r_g = 2Gm_{\text{out}} < r_\pm = Gm_{\text{in}} = \sqrt{G}e$, as it must be at $\Delta m < -(e^2+GM^2/(2\sqrt{G}e)$. It is impossible to draw differently, because $\sigma_{\text{out}}(\rho_0) = -1$.

Extreme Black Hole at $e^2 = Gm_{\text{in}}^2$: Finite Motion

Let us investigate the finite motion, when $\Delta m > -M$, $\sigma_{\text{in}}(0) = +1$ and, therefore, $\sigma_{\text{in}} = +1$ everywhere (see Fig. 4). Now we have only one turning point ρ_0, since the second root is negative. It is necessary to establish relations between three crucial radii of the shell: ρ_0, $\rho_{\sigma_{\text{in}}}$, $\rho_{\sigma_{\text{out}}}$ and also the radius of the double horizon $r_\pm$. For the turning point we find

$$\rho_0 = \frac{1+\frac{\Delta m}{M}}{(e+\sqrt{G}M)^2}, \tag{170}$$

It is simply to verify that $r_\pm < \rho_0 < \rho_{\sigma_{\text{in}}}$. Consequently, the shell, starting from $r = 0$, come out beyond the double horizon and meet the turning point, before the reaching the point $\rho_{\sigma_{\text{in}}}$ (the point of changing sign of σ). This means, that $\sigma_{\text{in}} = +1$ everywhere at the trajectory.

Now we investigate the point $\rho_{\sigma_{\rm out}}$. For the self-attractive shell with $e^2 < GM^2$, the point of the changing sign of $\sigma_{\rm out}$ does not exist ($\Delta m < 0$). Since $\sigma_{\rm out}(0) = -1$, then $\sigma_{\rm out} = -1$ everywhere at the trajectory. Conformal diagram for the corresponding internal metric is shown at the left panel in Fig. 25.

For the self-repulsive shell with $e^2 > GM^2$ there is point of changing sign of $\sigma_{\rm out}$, and $\sigma_{\rm out}(0) = +1$. For this reason the shell, in principle, may come out to the R_--, and also to the R_+–region of the external metric. In the first case we have $\rho_{\sigma_{\rm out}} < r_g = 2Gm_{\rm out}$, and, respectively, in the second case we have $\rho_{\sigma_{\rm out}} > \rho_0 > r_g = 2Gm_{\rm out}$.

Let us verify, at first, that $\rho_0 \geq r_g$:

$$\frac{(e+\sqrt{G}M)^2}{2(M-\Delta m)} > 2Gm_{\rm out} \quad \Rightarrow \quad (e-\sqrt{G}M+\sqrt{G}\Delta m)^2 \geq 0. \tag{171}$$

Consider, further, the case $\rho_{\sigma_{\rm out}} < r_g$, when turning point is placed in the R_-–region of the external metric. Well, we have

$$-\frac{e^2-GM^2}{2\Delta m} < 2Gm_{\rm out} \quad \Rightarrow \quad \sqrt{G}(2m_{\rm out}-m_{\rm in}) > -\frac{GM^2}{2}. \tag{172}$$

The expression at the left part of the last inequality may be as positive and negative. For this reason, we well the both cases separately.

(1) If

$$-\frac{GM^2}{2} < \sqrt{G}(2m_{\rm out}-m_{\rm in}) < 0, \tag{173}$$

then

$$-2m_{\rm out}+m_{\rm in})M \quad \Rightarrow \quad m_{\rm in} < 0 < 2m_{\rm out} \quad \Rightarrow \quad \rho_{\sigma_{\rm out}} < r_g. \tag{174}$$

(2) If

$$2\sqrt{G}\Delta m + e > 0, \tag{175}$$

then

$$2\sqrt{G}\Delta m + e > 0 \quad \Rightarrow \quad m_{\rm in} < 2m_{\rm out} < M+m_{\rm in} \quad \Rightarrow \quad \rho_{\sigma_{\rm out}} < 2r_g. \tag{176}$$

In result, for implementation of the inequality $\rho_{\sigma_{\rm out}} < 2r_g$, it is necessary

$$m_{\rm out} < \frac{M+m_{\rm in}}{2}. \tag{177}$$

The the moving shell is coming in the R_+–region, then $\rho_{\sigma_{\rm out}} > 2Gm_{\rm out}$ and $(m_{\rm in}2m_{\rm out})^2 > M^2$. In the case $m_{\rm in} > 2m_{\rm out}$, the inequality is valid, $m_{\rm in} - 2m_{\rm out} > M$, and, therefore, $2m_{\rm out} < m_{\rm in} - M < 0$. However, the last inequality contradicts to condition $m_{\rm out} > 0$. Meantime, if $m_{\rm in} < 2m_{\rm out}$, then

$$2m_{\rm out} - m_{\rm in} > M \quad \Rightarrow \quad m_{\rm out} > \frac{M + m_{\rm in}}{2}. \tag{178}$$

It is remained to verify that $\rho_0 < \rho_{\sigma_{\rm out}}$. Really, we obtain

$$\frac{(e+\sqrt{G}M)^2}{2(M-\Delta m)} < -\frac{e^2 - GM^2}{1\Delta m} \quad \Rightarrow \quad 2m_{\rm out} - m_{\rm in} > M. \tag{179}$$

The total conformal diagram at $m_{\rm in} > 2m_{\rm out}$ corresponds to the left panel in Fig. 27, and at $m_{\rm in} < 2m_{\rm out} < M + m_{\rm in}$ and $\Delta m < (M - m_{\rm in})/2$, respectively, to the right panel in Fig. 30. At last, the total conformal diagram at $2m_{\rm out} > M + m_{\rm in}$ and $\Delta m > (M - m_{\rm in})/2$ is shown in Fig. 38.

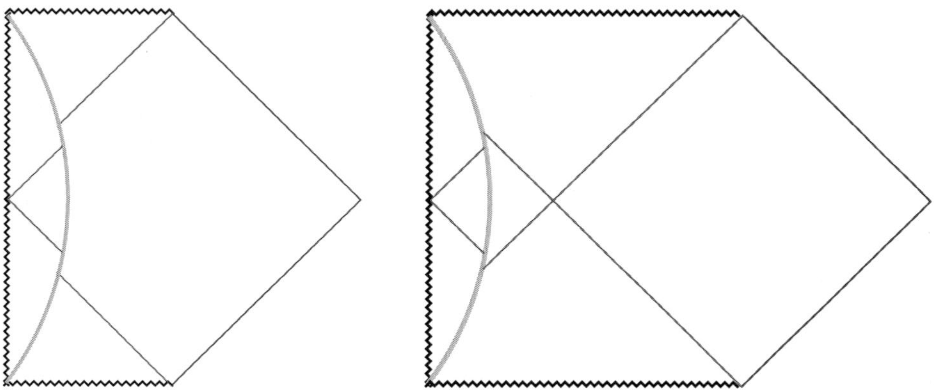

Figure 38. The case of extreme black hole. Finite motion at $2m_{\rm out} > M + m_{\rm in}$ and $\Delta m > (M - m_{\rm in})/2$.

Black Hole at $e^2 < Gm^2_{\rm in}$: Finite Motion

Now we initiate investigation of the most multivariant case with $\Delta m < 0$, when inside the neutralizing shell there is the reissner–Nordström black hole with

$e^2 < Gm_{\text{in}}^2$. The Carter–Penrose diagram for this internal metric is shown in Fig. 4.

Now there is the $T_{\pm}$–region, where the sign of σ_{in} may be changing, and two $R_{\pm}$–regions, where may be the turning points. All possibilities at $\Delta m < 0$ and $e^2 < Gm_{\text{in}}^2$ for the internal metric are shown in Fig. 39, and for the external metric, respectively, in Fig. 40

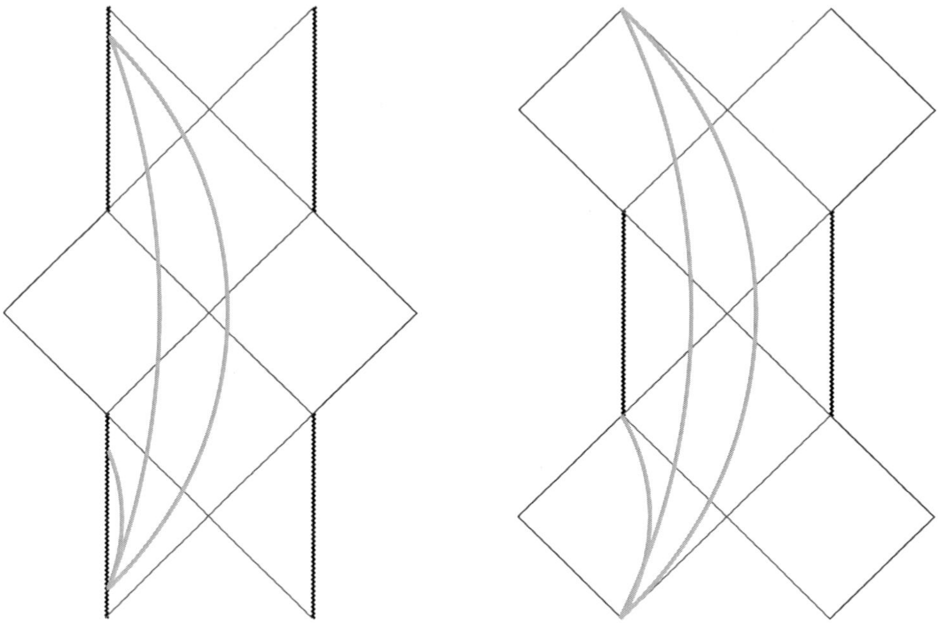

Figure 39. All possible trajectories of the shell at $\Delta m < 0$ and $e^2 < Gm_{\text{in}}^2$ for the internal metric.

Before to proceed further, let us see the graphs for the trinomial $A(\rho)$ in dependance of the relations between the specified parameters in Figs. 41–42. It is possible, in principle, the infinite motion at $\Delta m/M < -1$, parabolic motion at $\Delta m/M < -1$ and finite motion at $0 > \Delta m/M > -1$ with the one turning point. We see, that for every type of motion, the graphs in Figs. 41–42 differ in dependance of the sign of the coefficient B and the sign of the discriminant D.

Let us consider the behavior of B and D in variation of the value $\Delta m < 0$. Both B and D are the linear functions of $\Delta m < 0$, growing in the case of the self-repulsive shell ($e^2 > GM^2$), and decreasing for the self-attractive shell ($e^2 <$

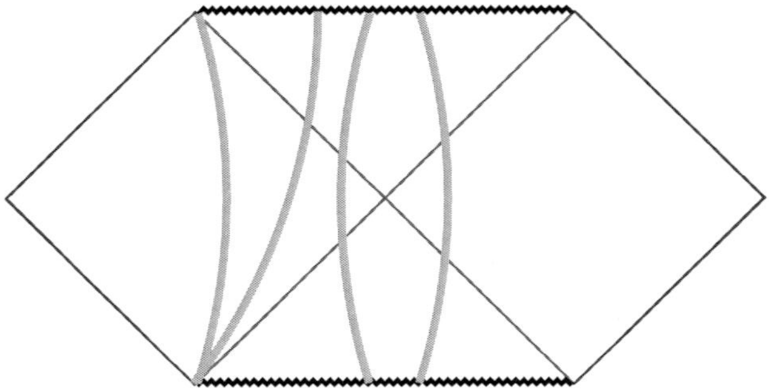

Figure 40. All possible trajectories of the shell at $\Delta m < 0$ and $e^2 < Gm_{\text{in}}^2$ for the external metric.

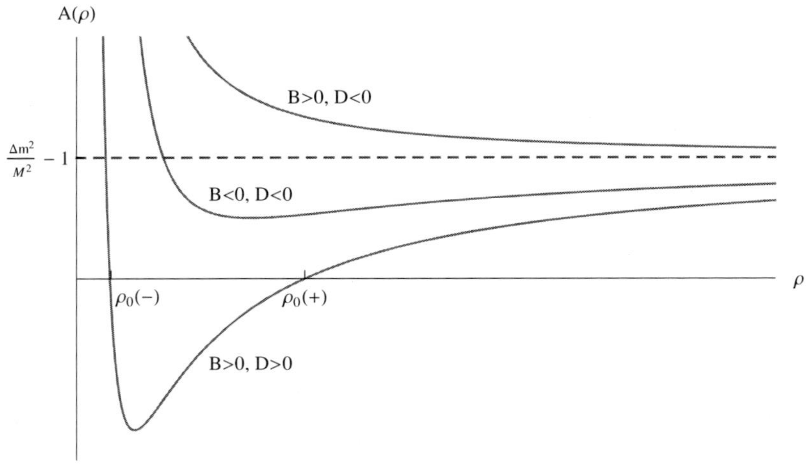

Figure 41. The graphs for the trinomial $A(\rho)$ for infinite motion, when $\Delta m/M < -1$. The allowed region for motion is at $A(\rho) \geq 0$.

GM^2). In the last case, since everywhere $B > 0$, the turning points are absent for the infinite motion (the both roots of equation $A = 0$ are negative).

Meantime, at the finite motion one root of equation $A = 0$ is positive and the

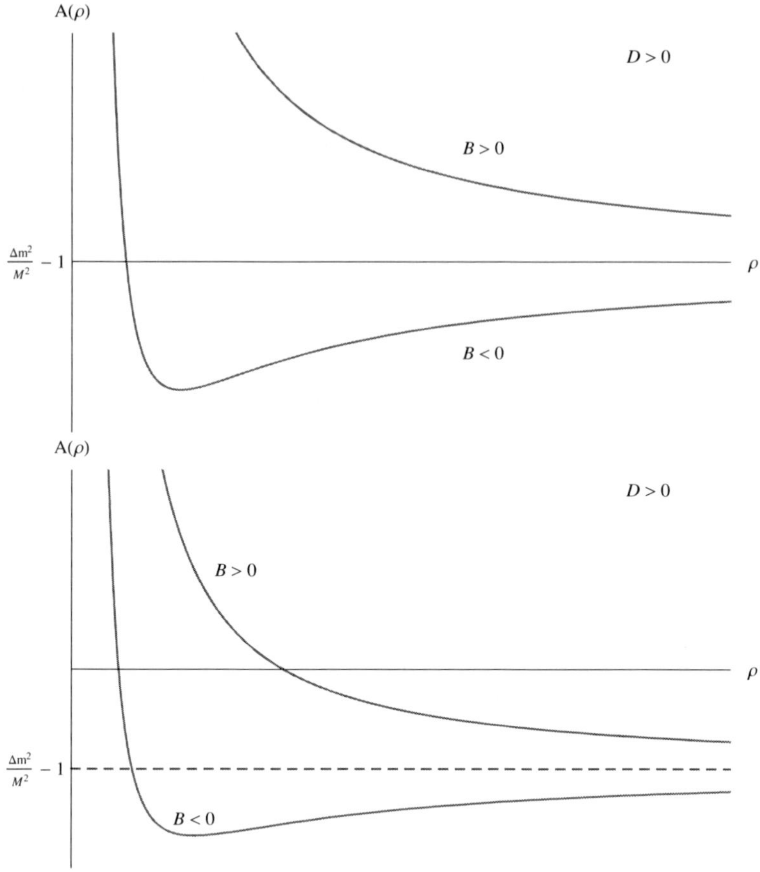

Figure 42. The graphs for the trinomial $A(\rho)$ for parabolic motion (the upper panel), when $\Delta m/M = -1$ and motion is possible only at $B > 0$, and, respectively, for finite motion (the lower panel), when $\Delta m/M > -1$ and there is only the one turning point.

other one is negative:

$$\frac{1}{2\rho_0 M} = \frac{-B + \sqrt{D}}{(e^2 - GM^2)^2}. \tag{180}$$

Note, that intersection of the curves on the graphs in Fig. 43 takes place at $\Delta m = -M$. For comparison on the graph is shown the case $e^2 = GM^2$, when $\sqrt{D} =$

$B = 2GMm_{\text{out}}$. For the self-repulsive shell two of these possibilities are realized, depending on the positive or negative value of $B(\Delta m)$ at he point $\Delta m = -M$. On the graph in Fig. 43 it is shown the case $B(-M) > 0$ and, for comparison, the strict line $e^2 = GM^2$, when $\sqrt{D} = B = 2GMm_{\text{out}}$.

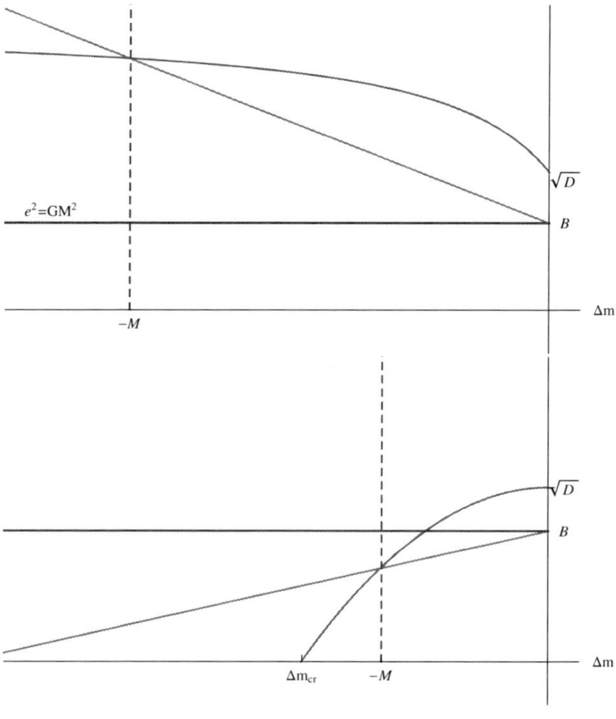

Figure 43. The case $e^2 = GM^2$, when $\sqrt{D} = B = 2GMm_{\text{out}}$ (a the upper panel). The case $B(-M) > 0$ and, for comparison, the strict line $e^2 = GM^2$, when $\sqrt{D} = B = 2GMm_{\text{out}}$ (at the lower panel).

Black Hole at $e^2 < Gm_{\text{in}}^2$: The Case $GM^2 < e^2 < GM^2 + 2GMm_{\text{out}}$

Condition $B(-M) > 0$, is held at if $GM^2 < e^2 < GM^2 + 2GMm_{\text{out}}$. From the point of view of the internal metric this case qualitatively is not different from the self-attractive shell: the infinite motion proceeds without the turning points ($B > 0$, $D > 0$ — The both roots re negative or complex, if $D < 0$), and in the case of finite motion there is one turning point, in which connection $\sqrt{D} >$

B. However, there is difference with the external metric: $\sigma_{\text{in}}(0) = +1$, but $\sigma_{\text{out}}(0) = -1$ for the self-repulsive shell, and $\sigma_{\text{out}}(0) = -1$ for the self-attractive shell. At the same time $\sigma_{\text{in}}(\infty) = \sigma_{\text{out}}(\infty) = -1$.

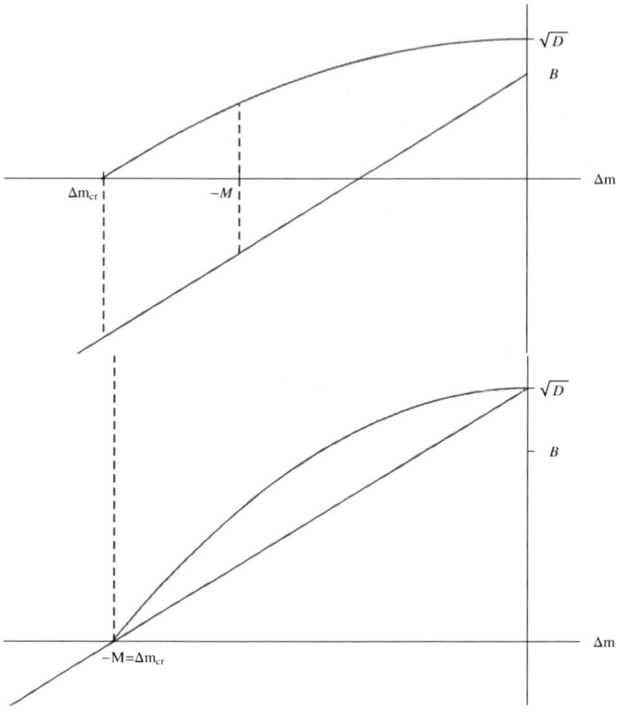

Figure 44. The case $e^2 > GM^2 + 2GMm_{\text{out}}$. The intersection points of the graphs $\sqrt{D(\Delta m)}$ and $B(\Delta m)$ with the vertical line $\Delta m = -M$ are placed symmetrically with respect to the horizontal axes (the upper pairs of graphs). The lower pairs of graphs corresponds to the transition case.

Black Hole at $e^2 < Gm_{\text{in}}^2$: The Case $e^2 > GM^2 + 2GMm_{\text{out}}$

If $e^2 > GM^2 + 2GMm_{\text{out}}$, then the graph has the for, shown at he upper panel in Fig. 44. The intersection points of the graphs $\sqrt{D(\Delta m)}$ and $B(\Delta m)$ with the vertical line $\Delta m = -M$ are placed symmetrically with respect to the horizontal axes.

For finite motion, as previously, the change of the sign B does not play any role, but situation now is quite the different for the infinite motion. If $\Delta m < \Delta m_{\rm cr}$, where

$$
\begin{aligned}
\Delta m_{\rm cr} &= -\frac{(GM^2-e^2)^2+4G^2m_{\rm out}^2}{4Gm_{\rm out}(e^2-GM^2)} \\
&= \frac{M}{2}\left(\frac{e^2-GM^2}{2GMm_{\rm out}}+\frac{2GMm_{\rm out}}{e^2-GM^2}\right) \leq M,
\end{aligned} \tag{181}
$$

(an equality is reached exactly at $e^2 = GM^2 + 2GMm_{\rm out}$), then turning point are absent. However at $\Delta m > \Delta m_{\rm cr}$ it appears the range of parameters, when $D > 0$, and $B < 0$, i. e., the equation $A = 0$ has two positive roots, $B^2 > D$. Since the shell is starting at infinity, then the turning point is the bigger root

$$
\frac{1}{2\rho_0 M} = \frac{-B-\sqrt{D}}{(e^2-GM^2)^2}. \tag{182}
$$

At $\Delta m \to -M$ this turning point is going away to infinity. Afterwards the bigger turning point becomes the negative root for the finite motion ($\Delta m > M$), and the smaller root becomes the turning point

$$
\frac{1}{2\rho_0 M} = \frac{-B+\sqrt{D}}{(e^2-GM^2)^2}. \tag{183}
$$

Black Hole at $e^2 < Gm_{\rm in}^2$: The Case $e^2 = GM^2 + 2GMm_{\rm out}$

At last, for comparison and clearness we demonstrate the graphs for the transition case, shown at he lower panel in Fig. 44.

6.4. Self-attractive Shell

For investigation the possible shell trajectories at $\Delta m < 0$, it is requested to consider the relations between the turning points (when they exist), the point of changing sign of $\sigma_{\rm in}$ and $\sigma_{\rm out}$, the radii of horizons $r_\pm$ and the gravitational radius of the external metric r_g.

it appears, that it is useful to fix Δm, since in this choice the points of changing sign of $\rho_{\rm in}$ and $\rho_{\rm out}$ are fixed. It is useful also to change the value of $m_{\rm out}$ (and simultaneously the value of $m_{\rm in}$), starting at $m_{\rm out} = 0$. For the case we writhe the corresponding equation:

$$\sigma_{\rm in}\sqrt{\dot\rho^2+F_{\rm in}}-\sigma_{\rm out}\sqrt{\dot\rho^2+F_{\rm out}}=\frac{GM}{\rho}, \tag{184}$$

where

$$F_{\rm in}=1-\frac{2Gm_{\rm in}}{\rho}+\frac{Ge^2}{\rho^2},\quad F_{\rm out}=1-\frac{2Gm_{\rm out}}{\rho}. \tag{185}$$

The square of Israel equation gives

$$\frac{\Delta m}{M}\left(1-\frac{\rho_{\sigma_{\rm in}}}{\rho}\right)=\sigma_{\rm in}\sqrt{\dot\rho^2+F_{\rm in}} \tag{186}$$

$$\frac{\Delta m}{M}\left(1-\frac{\rho_{\sigma_{\rm out}}}{\rho}\right)=\sigma_{\rm out}\sqrt{\dot\rho^2+F_{\rm out}} \tag{187}$$

From this it follows, in particular, that if the turning point ($\dot\rho=0$) coincides with the one of horizons ($F=0$), then this point coincides with point of the changing sign of σ.

From equation for the turning point (147) it follows the conditions

$$\dot\rho^2=A=\frac{(e^2-GM^2)^2}{4\rho^2M^2}+\frac{2B}{2\rho M}+\frac{\Delta m^2}{M^2}-1\geq 0. \tag{188}$$

For the turning point $\rho=\rho_0$, where $A=0$, we find with the help of (93):

$$D=(e^2-GM^2)^2+4GM\Delta m(e^2-GM^2)m_{\rm out}+4G^2M^2m_{\rm out}^2. \tag{189}$$

The permissible regions for the possible shell motions $A(\rho)\geq 0$ are shown in Figs. 41 and 42. We see, that behavior of the function $A(\rho)$, defining the turning point (or their absence), is qualitatively different for self-attractive, $e^2<GM^2$, and self-repulsive, $e^2>GM^2$, shells. Also, in the first case the function $B(m_{\rm out})$ is always positive (at $\Delta m<0$), but in the second case the function $B(m_{\rm out})$ is initially negative ($\Delta m=0$), and afterwards becomes the positive. Besides, the radius $\rho_{\sigma_{\rm out}}$ exists, only if $e^2>GM^2$. For this reason, all these cases must be analyzed separately.

We start from the relatively simple of the self-attractive shell, $e^2<GM^2$, which is intuitively most easily understood, since there is limit $e^2\to 0$. In this case $0<\rho<\infty$, $\rho_{\sigma_{\rm in}}(-\infty)=-1$, $0<\rho<\infty$ and $\rho_{\sigma_{\rm in}}(0)=+1$. Under these conditions the shell is inevitably comes through the point $\rho_{\sigma_{\rm in}}$, because everywhere at the trajectory $\sigma_{\rm out}=-1$. We need to verify the validity of inequalities

$r_- < \rho_{\sigma_{in}} < r_+$, i. e., the placement of the point $\rho_{\sigma_{in}}$ in the T–region. To do this we consider the relations in the T–region of the internal metric:

$$F_{in} = 1 - \frac{2Gm_{in}}{\rho} + \frac{Ge^2}{\rho^2} < 0, \tag{190}$$

It is enough to show, that $F_{in}(\rho_{\sigma_{in}}) < 0$. At first we prove the validity of the required inequality for the limiting relation $e^2 = GM^2$. We have

$$F_{in}(\rho_{\sigma_{in}}) = \frac{1}{M^2}[m^2_{out} - (m_{in} - M)((m_{in} - M))] < 0, \tag{191}$$

since from the relation $\Delta m < -M$ it follows that $0 < m_{out} < m_{in} - M$, which was to be proved. Secondly, at the fixed values of M, m_{in} and Δm, we will diminish the charge e. Herewith $\rho_{\sigma_{in}}$ is diminishing, but r_+ is growing, i. e., the inequality $\rho_{\sigma_{in}} < r_+$ is conserved. It remains to prove that the second inequality $r_- < \rho_{\sigma_{in}}$ is also conserved, when both $\rho_{\sigma_{in}}$ and r_+ are diminishing with the diminishing of the charge e. The function $F_{in}(\rho)$ reaches the minimum$F_{in}(\min) = 1 - Gm^2_{in}/e^2 < 0$ at $\rho_{\min} = e^2/m_{in}$. Prove now, that always $\rho_{\sigma_{in}} > \rho_{\min}$. It is really

$$\rho_{\sigma_{in}} > \rho_{\min} \Rightarrow e^2 + GM^2 > \frac{2\Delta m e^2}{m_{in}} \Rightarrow GM^2 > e^2 - 2\frac{m_{in}}{m_{in}}e^2, \tag{192}$$

$GM^2 > e^2$, wherefrom it follows the requested result. It is evident, in the case of the self-attractive shell always $D > 0$, and $\sigma_{out} = -1$ everywhere at trajectory, since $\rho_{\sigma_{out}}$ does not exist ($\rho_{\sigma_{out}} < 0$), and $\sigma_{out}(0) = \sigma_{out}(\infty) = -1$. The following relations are valid:

$$\Delta m = m_{out} - m_{in}, \quad m_{out} > 0, \quad \Delta m > -m_{in}, \tag{193}$$

$$B_{\min} = -\frac{\Delta m}{M}(GM^2 - e^2), \ \sqrt{D_{\min}} = GM^2 - e^2, \ \rho_0 = \frac{GM^2 - e^2}{2(\pm M + \Delta m)}. \tag{194}$$

It is clear, that there is the turning point $\rho_0 > 0$ at $M > -\Delta m$ (by fixing the sign '+' in), corresponding to the finite motion $0 \le \rho \le \rho_0$ at $M < -\Delta m$ when is realized the infinite motion without the turning point, $0 \le \rho \le \infty$. At infinite motion we have the following chain of inequalities:

$$e^2 \le GM^2 \le G\Delta m^2 \le Gm^2_{in}, \quad m_{out} \ge 0. \tag{195}$$

At finite motion there are two separate cases. In the first case

$$e^2 \leq G\Delta m^2 \leq GM^2, \quad m_{\rm out} \geq 0. \tag{196}$$

and $\Delta m^2 \leq M^2$ at $m_{\rm out} = 0$. Then, under the increasing of $m_{\rm out}$, the mass $m_{\rm out}$ happens to be greater than M. Respectively, in the second case $G\Delta m^2 \leq e^2 \leq GM^2$. Therefore, we cannot start with $m_{\rm out} = 0$, but only with

$$m_{\rm out,min} = \frac{|e|}{\sqrt{G}} + \Delta m > 0. \tag{197}$$

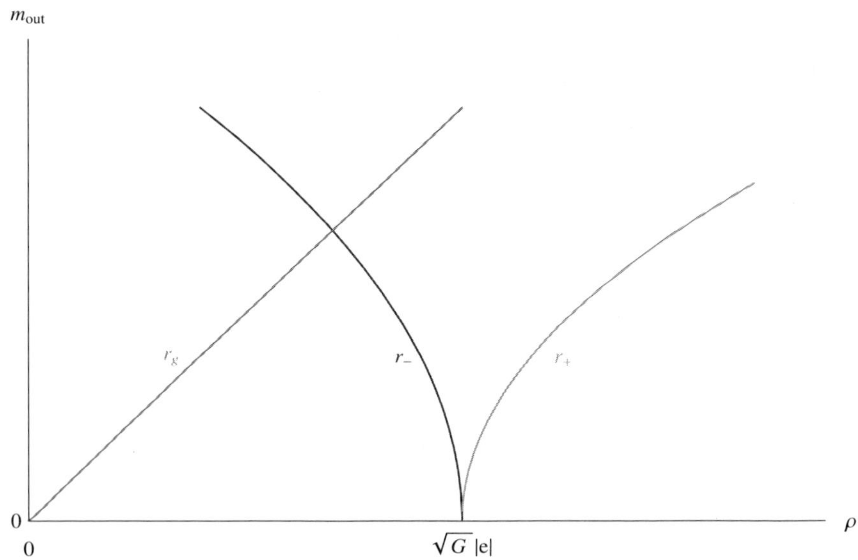

Figure 45. The relative position of radii $r_\pm$ and r_g in dependance of $m_{\rm out}$ in the case of the self-attractive shell.

Remind, that $\rho_{\sigma_{\rm in}} = const$, $\rho_{\sigma_{\rm out}}$ do not exist, and the horizons $r_\pm$ of the internal metric with growing value of $m_{\rm in}(m_{\rm out})$ are changed in a following way:

$$r_\pm = Gm_{\rm in} \pm \sqrt{G^2 m_{\rm in}^2 - Ge^2}, \tag{198}$$

$$\frac{\partial r_\pm}{\partial Gm_{\rm in}} = \frac{r_\pm}{\sqrt{G^2 m_{\rm in}^2 - Ge^2}}. \tag{199}$$

The relative position of radii $r_\pm$ and r_g in dependance of $m_{\rm out}$ is shown in Fig. 45.

Let us clarify now, where is placed the point $\rho_{\sigma_{\rm in}}$. To do this we verify the validity of inequality $\rho_{\sigma_{\rm in}} > r_\pm$ at the point, when $m_{\rm out} = 0$ and $\Delta m = m_{\rm in}$:

$$GM > -G\Delta m \pm \sqrt{G^2\Delta m^2 - Ge^2} = r_\pm, \tag{200}$$

$$G(M+\Delta m) > \sqrt{G^2\Delta m^2 - Ge^2}. \tag{201}$$

It follows directly from here, that at the infinite motion, when $m_{\rm out} = 0$ and $M+\Delta m < 0$, the following relations are held

$$\rho_{\sigma_{\rm in}} < r_{+,min} < r_+, \tag{202}$$

Respectively, at the finite motion, when $m_{\rm out} = 0$ and $M+\Delta m > 0$, the following relations are held

$$\rho_{\sigma_{\rm in}} > r_{-,\max} > r_-. \tag{203}$$

We demonstrate now that at the infinite motion $\rho_{\sigma_{\rm in}} > r_-$. It must be held the following inequality

$$\sqrt{G^2\Delta m^2 - Ge^2} > G(M+\Delta m). \tag{204}$$

Since $\Delta m < -M$, we have

$$GM^2 + 2GM\Delta m < -GM^2 < -e^2. \tag{205}$$

It is proved in a similar way, that at the finite motion $\rho_{\sigma_{\rm in}} < r_{+,min}$. At this step it is clear, that at the infinite motion there are no obstacles for shell to move from infinity to $r = 0$, at $\sigma_{\rm in}(\infty) = -1$ and $\sigma_{\rm in}(0) = +1$, since $\rho_{\sigma_{\rm in}}$, it is proved, is placed in the T–region between the horizons. In other words, for the self-gravitating shell with $e^2 < GM^2$ the only condition for the infinite motion is $\Delta m < -M$. The Carter–Penrose diagram for infinite motion of the self-gravitating shell at $\Delta m < 0$, $e^2 < Gm^2_{\rm in}$ coincides with diagram in Fig. 46.

Let us see now, what is happened at the finite motion. Now we have the turning point at ρ_0. The question is, where this point is placed with respect to $r_\pm$, and what is the sign of $\sigma(\rho_0)$. We already know, that there are two different cases. In the first case, when $e/\sqrt{G} < -\Delta m < M$, we have

$$m_{\rm out,min} = 0, \quad \rho_{0,\rm min} = \frac{GM^2 - e^2}{2(M - m_{\rm in})}. \tag{206}$$

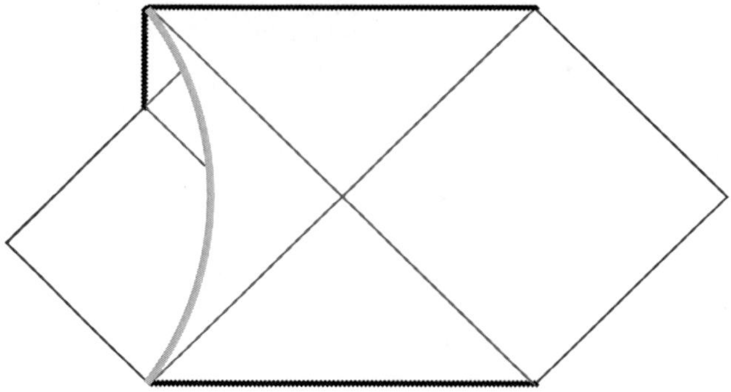

Figure 46. The complete conformal Carter–Penrose for infinite motion of the self-gravitating shell.

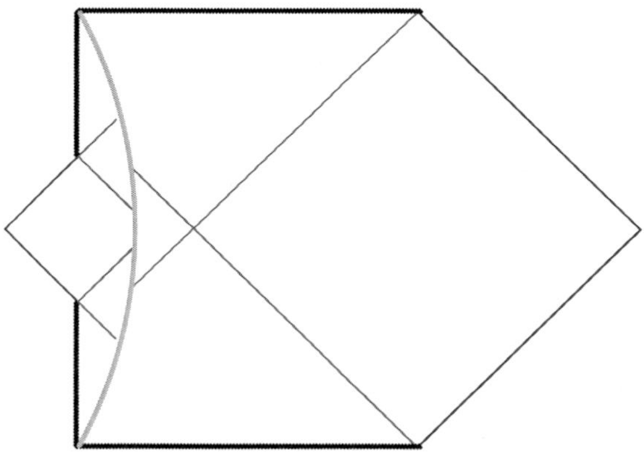

Figure 47. The Carter–Penrose diagram for the case of finite motion of the self-attractive shell at $\Delta m < 0$, $e^2 < Gm_{\text{in}}^2$.

Let us demonstrate, that $\rho_{0,\min} \geq r_{+,\min}$. Really, the inequality

$$\rho_{0,\min} = \frac{GM^2 - e^2}{2(M - m_{\text{in}})} > Gm_{\text{in}} + \sqrt{G^2 m_{\text{in}}^2 - Ge^2} \tag{207}$$

is transformed to the evident inequality

$$r_+^2 - 2GMr_+ + G^2M^2 = (r_+ + GM)^2 \geq 0. \tag{208}$$

This means, that $\sigma_{\rm in}(\rho_{0,\rm min}) = -1$, and the shell, starting from the zero radius, moves through the T–region into the R–region of the internal netric. With increasing of $m_{\rm out}$, there are also increased both the ρ_0 and r_+, buy, however, $\rho_{\sigma_{\rm in}}$ is not changed. Therefore the sign of σ is inevitably changing at the trajectory. This is confirmed also by the fact, that the coinciding of ρ_0 and r_+ means simultaneously the intersection at this point also with $\rho_{\sigma_{\rm in}}$, but, however, $\rho_{\sigma_{\rm in}} < r_+$. The Carter–Penrose diagram for the case of finite motion of the self-attractive shell at $\Delta m < 0$, $e^2 < Gm^2_{\rm in}$ is shown in Fig. 47 and coinside with the right diagram in Fig. 31.

It is remained to consider the second case, when $-\Delta m < e/\sqrt{G} < M$ and the minimal value of $m_{\rm out}$ is already nonzero:

$$m_{\rm out,min} = \frac{e}{\sqrt{G}} + \Delta m > 0, \quad m_{\rm in,min} = \frac{e}{\sqrt{G}}. \tag{209}$$

It is not difficult to show, that $\rho_{\sigma_{\rm in}} > e/\sqrt{G} = r_{-,\rm max} = r_{+,\rm min}$. Really, the inequality

$$\frac{GM^2 - e^2}{2\Delta m} > e\sqrt{G} \tag{210}$$

transforms to the evident one

$$(e + \sqrt{G}\Delta m)^2 + G(M^2 - \Delta m^2) > 0. \tag{211}$$

Let us calculate $\rho_{0,\rm min}$ in this case. We have:

$$B_{\rm min} = \frac{\Delta m}{M}(e^2 + GM^2) + 2\sqrt{G}Me, \tag{212}$$

$$\sqrt{D_{\rm min}} = e^2 + GM^2 + 2\sqrt{G}\Delta me > 0, \tag{213}$$

$$\rho_{0,\rm min} = \frac{(e + \sqrt{G}M)^2}{2(M - \Delta m)}. \tag{214}$$

By direct calculations it is easily to verify, that $\rho_{0,\rm min} < \rho_{\sigma_{\rm in}}$ (see Fig. 48). Therefore, $0 < \sqrt{G}e < r_0 < \rho_{\sigma_{\rm in}}$ at $m_{\rm out} = m_{\rm out,min}$.

With the growing of $m_{\rm out}$ (ad $m_{\rm in}$) the radius r_+ inevitably intersects $\rho_{\sigma_{\rm in}}$, i. e., $\rho_{\sigma_{\rm in}}$ is inevitably coming into the T–region. At the same point ρ_0 intersects $\rho_{\sigma_{\rm in}}$ and touches with r_+. This means, that ρ_0 from R_+–region beyond the r_+ goes into R_-–region beyond the r_+. This event proceeds at the critical value of $m_{\rm in} = m_{\rm in,cr}$, where

$$m_{\rm in,cr} = -\frac{e}{2\sqrt{G}}\left(\frac{e^2 + GM^2}{2\sqrt{G}e\Delta m} + \frac{2\sqrt{G}e\Delta m}{e^2 + GM^2}\right) > \frac{e}{\sqrt{G}}. \tag{215}$$

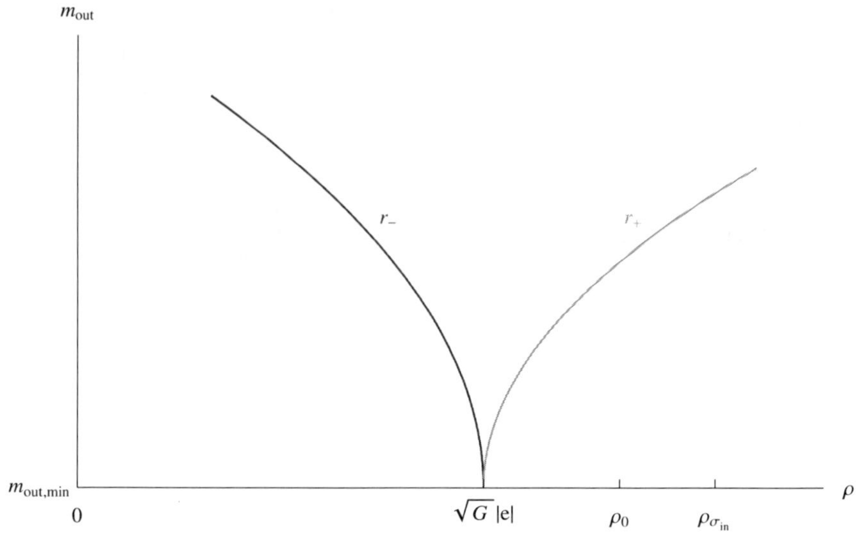

Figure 48. The mutual positional relationship of the characteristic radii $0 < \sqrt{G}e < r_0 < \rho_{\sigma_{in}}$ and $r_\pm$.

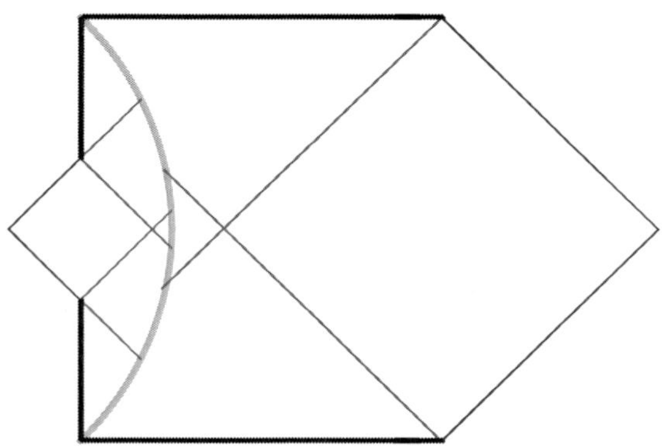

Figure 49. The Carter–Penrose diagram for the case of finite motion of the self-attractive shell at $-\Delta m < \sqrt{G}e < M$ and $m_{in} > m_{in,cr}$.

The corresponding Carter–Penrose diagram for the case of finite motion of the self-attractive shell at $-\Delta m < \sqrt{G}e < M$ and $m_{\rm in} > m_{\rm in,cr}$ is shown in Fig. 49 and coincides with the left diagram in Fig. 31.

6.5. Self-repulsive Shell

In the case of the self-repulsive shell, since $GM^2 < e^2 < Gm_{\rm in}^2$, then $M < m_{\rm in}$, and form condition $Gm_{\rm out} = m_{\rm in} + \Delta m \geq 0$ it follows, that $\Delta m \geq -m_{\rm in}$. At $m_{\rm out} = 0$ we obtain $\Delta m \geq -m_{\rm in} < -M$. At the same time, this means that at $m_{\rm out} = 0$ we start our analysis from the condition $|\Delta m| \geq M$, when

$$B_{\rm min} = \frac{\Delta m}{M}(e^2 - GM^2) < 0, \quad \sqrt{D_{\rm min}} = e^2 - GM^2 > 0, \tag{216}$$

We already know that in this case there are two turning points:

$$\rho_{0,\pm} = \frac{B \pm \sqrt{D}}{2M\left(1 - \frac{\Delta m^2}{M^2}\right)} = \frac{e^2 - GM^2}{2(-\Delta m \pm M)}, \tag{217}$$

in which connection $\rho_{0,+} < \rho_{0,-}$. It is evident, that $\rho_{0,+} < \rho_{\sigma_{\rm out}} < \rho_{0,-}$. We demonstrate below, that $\rho_{0,-} \leq r_-$. Really, this inequality easily transforms to the the evident one, $(GM - r_-)^2 \geq 0$.

As regards the relative placement of $\rho_{\sigma_{\rm in}}$, it only may say, that $\rho_{\sigma_{\rm out}} < \rho_{\sigma_{\rm in}} < r_+$. The first inequality is evident, and let us prove the second one:

$$\frac{e^2 + GM^2}{2m_{\rm in}} < Gm_{\rm in} + \sqrt{G^2 m_{\rm in}^2 - Ge^2} \Rightarrow \tag{218}$$

$$e^2 + GM^2 < 2Gm_{\rm in}^2 + 2m_{\rm in}\sqrt{G^2 m_{\rm in}^2 - Ge^2} \Rightarrow \tag{219}$$

$$G^2 M^2 < r_+^2 \Rightarrow GM < r_+. \tag{220}$$

Since $-\Delta m = m_{\rm in} > M$ at $m_{\rm out} = 0$, than it is equitable the same chain of inequalities, but with the change from r_+ to r_-. In result, we obtain, that $\rho_{\sigma_{\rm in}} < r_-$, if $GM < \rho_{\sigma_{\rm in}}$, and, consequently, $\rho_{\sigma_{\rm in}} > r_-$, if $GM > \rho_{\sigma_{\rm in}}$. Note, that inequality $GM < \rho_{\sigma_{\rm in}}$ is equivalent to the inequality $\rho_{\sigma_{\rm in}} < \rho_{0,-}$. In this case we have the following relation between the characteristic parameters: $GM < \rho_{\sigma_{\rm in}} < \rho_{0,-} < r_-$.

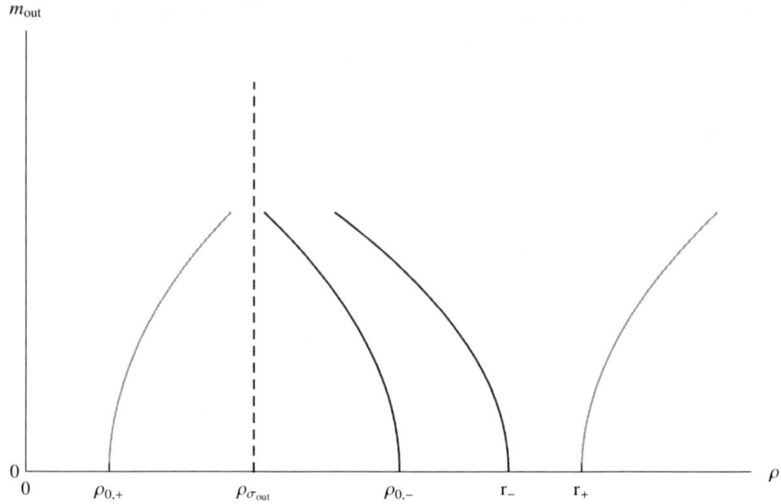

Figure 50. The relative positions of the characteristic radii in dependance of m_{out}.

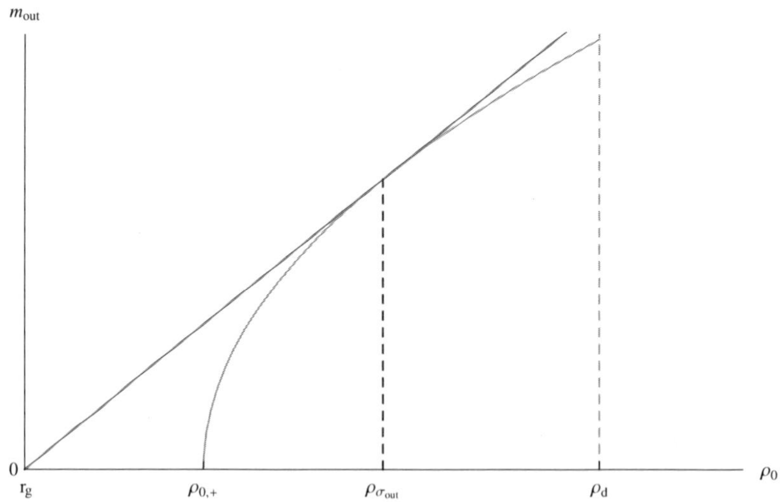

Figure 51. The relative positions of the characteristic radii in dependance of m_{out}.

What happens during the growing of $m_{\rm out}$? We fix the value of $\rho_{\sigma_{\rm out}}$ for subject to agreement. As we know, the radius of the inner horizon r_- is decreasing up to zero, and radius of the outer horizon r_+ is growing to infinity. Further, we have

$$\frac{\partial\rho_{0,-}}{\partial(2Gm_{\rm out})} = \frac{1 \pm \frac{B}{\sqrt{D}}}{2M\left(1 - \frac{\Delta m^2}{M^2}\right)} \frac{\partial B}{\partial(2Gm_{\rm out})} = \pm\frac{\rho_{0,\pm}}{\sqrt{D}}, \tag{221}$$

i. e., $\rho_{0,+}$ is growing, and $\rho_{0,-}$ is diminishing. Ultimately, they coalesce, whenthe discriminant $D = 0$, but B is still negative. It happens at

$$m_{\rm out} = m_{\rm out,1} = \frac{e^2 - GM^2}{2GM}\left(-\frac{\Delta m}{M} - \sqrt{\frac{\Delta m^2}{M^2} - 1}\right). \tag{222}$$

The double root equals

$$\rho_d = \frac{e^2 - GM^2}{2\sqrt{\delta m^2 - M^2}} > \rho_{\rm out}. \tag{223}$$

At the further growing of $m_{\rm out}$ the real roots of equation $A = 0$ are both disappeared, and discriminant is diminished to the minimal value, when $B = 0$, at

$$m_{\rm out} = -\frac{\Delta m}{M}\frac{e^2 - GM^2}{2GM}. \tag{224}$$

Then, the discriminant is growing again up to the zero at

$$m_{\rm out,2} = \frac{e^2 - GM^2}{2GM}\left(-\frac{\Delta m}{M} + \sqrt{\frac{\Delta m^2}{M^2} - 1}\right). \tag{225}$$

At the further growing of the $m_{\rm out}$, the two real roots of equation $A = 0$ are appearing again, but now both of them are negative. This means that the motion of the shell occurs without the turning points. The relative positions of the characteristic radii is shown in Figs. 50 and 51.

We start investigation from the region $0 < m_{\rm out} < m_{\rm out,1}$, when there are two turning points, separately for the finite motion at $0 < \rho_0 \le \rho_{0,+} \le r_g$, and for the infinite motion at $\rho_d \le \rho_0 \le \rho_{0,-}$ and $\rho_0 \le \rho < \infty$.

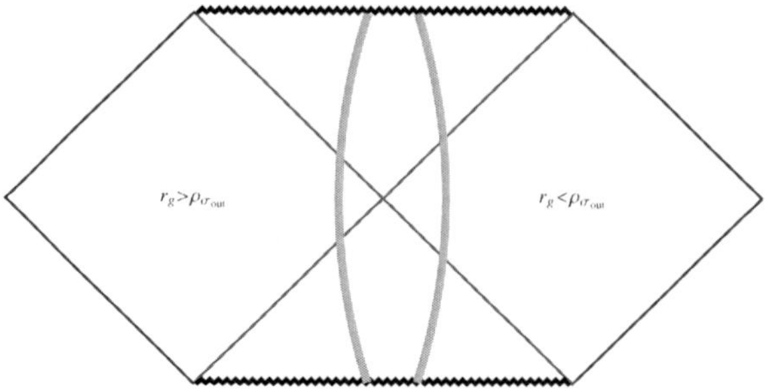

Figure 52. The possible conformal diagrams for the external Schwarzschild metric at $r_g \gtrless \rho_{\sigma_{\text{out}}}$.

Finite Motion with Turning Point

Let us describe the finite motion. We know, that at $m_{\text{out}} = 0$ the following relations are held: $0 = r_g < \rho_{0,+} \leq \rho_{\sigma_{\text{out}}}$. It is clear, that now $\rho_d > \rho_{\sigma_{\text{out}}}$. this mean, that with the growing of m_{out}, the curve ρ_0 intersects the line $\rho_{\sigma_{\text{out}}}$, i. e., at first $\sigma_{\text{out}}(\rho_0) = +1$, and then $\sigma_{\text{out}}(\rho_0) = -1$. It is also $\rho_0 = r_g$ at the intersection point, and at the further growing of m_{out}, it must be $\rho_0 > r_g$ (the trajectory turns out in the R–region of the external metric). Verify, that this is really takes place for ρ_d:

$$\rho_d > r_g \qquad \Rightarrow \qquad \frac{e^2 - GM^2}{2\sqrt{\Delta m^2 - M^2}} > 2Gm_{\text{out}} \qquad \Rightarrow \tag{226}$$

$$\left(\frac{\Delta m}{M} + \sqrt{\frac{\Delta m^2}{M^2} - 1} \right)^2 > 0. \tag{227}$$

The possible conformal diagrams for the external Schwarzschild metric are shown in Fig. 52.

To look into the internal metric, it is requested to find the place of $\rho_{\sigma_{\text{in}}}$. it is evident, that $\rho_{\sigma_{\text{in}}} > \rho_{\sigma_{\text{out}}}$. Find at first the condition, when $\rho_{\sigma_{\text{in}}} < r_-$ at m_{out}, i. e., at $m_{\text{in}} = \Delta m$:

$$\rho_{\sigma_{\text{in}}} \; < \; r_- \Rightarrow \frac{e^2 + GM^2}{2m_{\text{in}}} < Gm_{\text{in}} - \sqrt{G^2 m_{\text{in}}^2 - Ge^2} \Rightarrow \tag{228}$$

$$GM \; < \; r_- \Rightarrow GM < \rho_{\sigma_{\text{in}}}. \tag{229}$$

It appears, that at $GM < \rho_{\sigma_{\rm in}}$ simultaneously $\rho_{\sigma_{\rm in}} < r_-$ at $m_{\rm out} = 0$. With the growing of $m_{\rm out}$ the radius $\rho_{\sigma_{\rm in}}$ is standing at the same place, and, consequently, ρ_0 is growing, but r_- is diminishing. What are their relative values at the point ρ_d? We prove, that $\rho_d < \rho_{\sigma_{\rm in}}$:

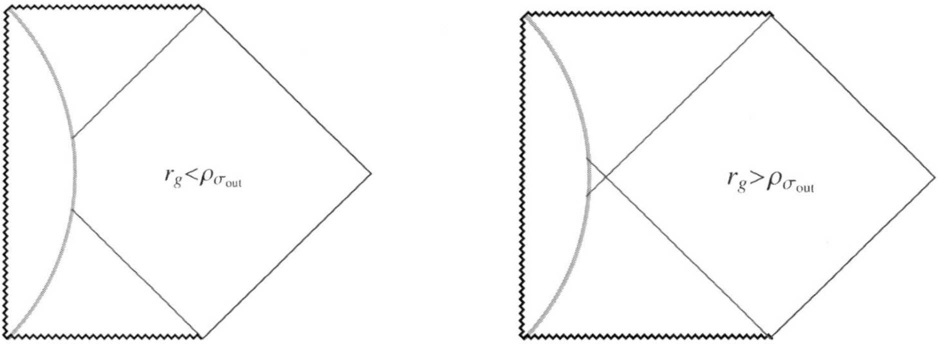

Figure 53. The Carter–Penrose diagrams for finite motion of the shell at $r_g \gtrless \rho_{\rm out}$.

$$\frac{e^2 - GM^2}{2\sqrt{\Delta m^2 - M^2}} < -\frac{e^2 + GM^2}{2\Delta m} \Rightarrow \rho_{\sigma_{\rm in}} < \sqrt{G}e. \tag{230}$$

However, in the described case this is valid from the outset (at $m_{\rm out} = 0$), if only $GM < \rho_{\sigma_{\rm in}}$. Meantime, if $GM > \rho_{\sigma_{\rm in}}$, then $\rho_{\sigma_{\rm in}} > r_-$ at $m_{\rm out} = 0$, and this inequality is held with the growing $m_{\rm out}$ (since r_- is diminishing). Since the turning point $\rho_{0,-} < r_-$ at $m_{\rm out} = 0$, then, $\rho_{0,-} < \rho_{\sigma_{\rm in}}$ at $GM > \rho_{\sigma_{\rm in}}$. Additionally, since $\rho_{0,-}$ is diminishing up to $\rho_0 = \rho_d$, then $\rho_d < \rho_{\sigma_{\rm in}}$. In result, our self-gravitating is expanded under the finite motion at $\Delta m < -M$, starting from $\rho = 0$, up to the turning point, which is smaller of the inner horizon r_- and without the intersection of $\rho_{\sigma_{\rm in}}$, i. e., $\sigma_{\rm in}(\rho_0) = +1$. The corresponding Carter–Penrose diagrams for finite motion of the shell at $r_g \gtrless \rho_{\rm out}$ are shown in Fig. 53.

Infinite Motion with Turning Point

Now we move to the infinite motion with the turning point $\rho_{0,-}$. The motion of the shell proceeds in the radial interval $\rho_{0,-} \le \rho < \infty$. At $m_{\rm out} = 0$ the relation

$\rho_{\sigma_{\rm out}} < \rho_{0,-}$ is held, and afterwards $\rho_{0,-}$ is diminishing up to ρ_d. Since, as it was before, $\rho_d > r_g$ and $\rho_d > \rho_{\sigma_{\rm out}}$, then in the external metric the motion takes place in the R_-–region and $\sigma_{\rm out}(\rho_0) = -1$. In the internal metric the the of the shell trajectory depends on the relation between $r_\pm$, $\rho_{\sigma_{\rm in}}$ and $\rho_{0,-}$. We already know, that at $m_{\rm out} = 0$ we have $\rho_{0,-} < r_-$. Respectively, for $\rho_{\sigma_{\rm in}}$ we have, $\rho_{0,-} < r_-$ at $GM < \rho_{\sigma_{\rm in}}$ and $\rho_{0,-} < r_- < \rho_{\sigma_{\rm in}} < r_+$ at $GM > \rho_{\sigma_{\rm in}}$.

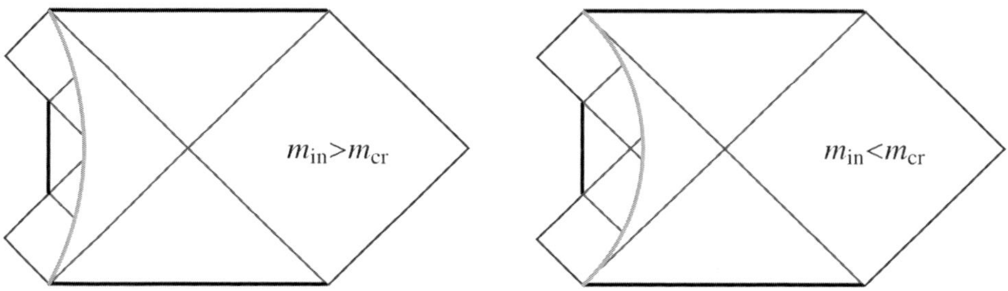

Figure 54. The Carter–Penrose diagrams for infinite motion of the self-attractive shell at $m_{\rm in} \gtrless m_{\rm in,cr}$.

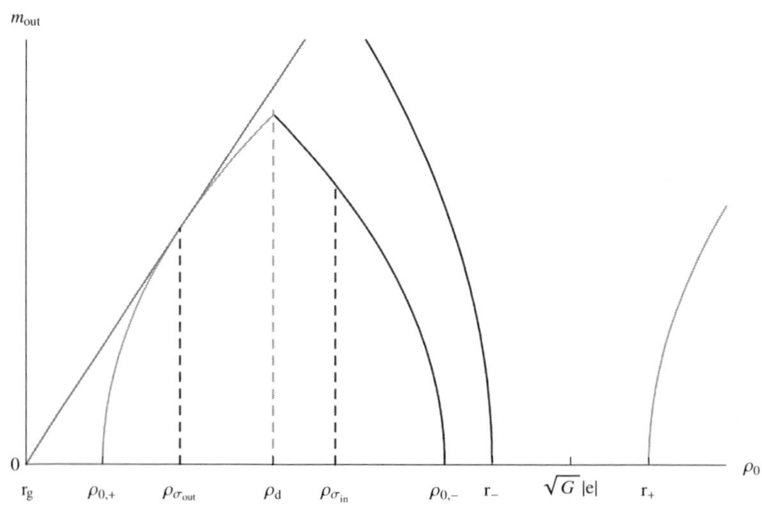

Figure 55. The relative displacement of r_g, $r_\pm$, ρ_0, $\rho_{\sigma_{\rm in}}$, $\rho_{\sigma_{\rm out}}$ and ρ_0 in dependance of $m_{\rm out}$ for infinite motion of the self-attractive shell.

Let us consider at first the last case, which is more simpler. It is clear, that the shell, contracting from the infinity, where $\sigma_{\rm in}(\infty)$, comes through the T_+–region between the horizons of the internal metric $r_\pm$, where the sign of $\sigma_{\rm in}$ is changing, and reach afterward theturning point $\rho_{0,-}$ in R_+–region near the central singularity. It happens at $0 \le m_{\rm out} < m_{\rm out,1}$, since $\rho_{0,-}$ is diminishing with the growth of $m_{\rm out}$, while $\rho_{\sigma_{\rm in}}$ remains at the same place. The corresponding Carter–Penrose diagram is shown at the left panel in Fig. 54.

Meanwhile, if $GM < \rho_{\sigma_{\rm in}}$, then at $m_{\rm out} = 0$ the moving shell does not meet at the trajectory the turning point $\rho_{\sigma_{\rm in}}$, i. e., $\sigma_{\rm in}(\rho_0) = -1$, and the turning point is placed in R_-–region near the central singularity. Since $\rho_d < \rho_{\sigma_{\rm in}}$ (see the text above), then with growing of $m_{\rm out}$ inevitably takes place the meeting: $\rho_{\sigma_{\rm in}} = r_- = \rho_{0,-}$. It happens at

$$m_{\rm in} = m_{\rm in,cr} = \frac{e}{2\sqrt{G}}\left(\frac{\rho_{\sigma_{\rm in}}}{\sqrt{G}e} + \frac{\sqrt{G}e}{\rho_{\sigma_{\rm in}}}\right) > M. \qquad (231)$$

Afterwards, $\rho_{0,-}$ becomes smaller than $\rho_{\sigma_{\rm in}}$, i. e., $\rho_{0,-} < \rho_{\sigma_{\rm in}} < r_-$ and therefore $\sigma_{\rm in}(\rho_0) = +1$, as in the case of $M > \rho_{\sigma_{\rm in}}$. The corresponding Carter–Penrose diagram is shown at the right panel in Fig. 54.

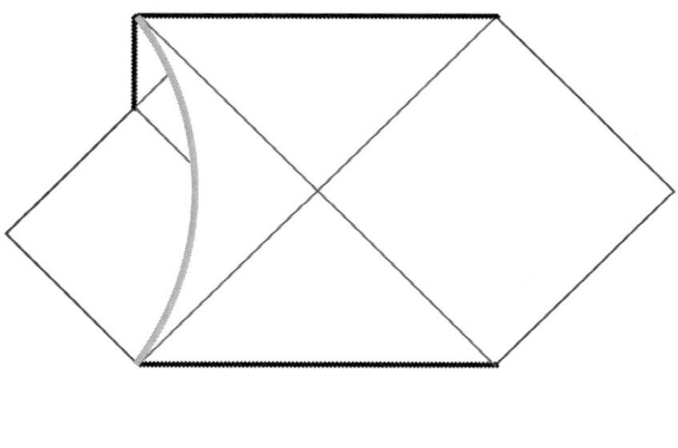

Figure 56. The Carter–Penrose diagram for infinite motion of the self-attractive shell in the case $m_{\rm out} > m_{\rm out,1}$.

At a further growing of $m_{\rm out} > m_{\rm out,1}$ we transfer to the region, where there are no turning points for the case of $\Delta m < -M$. Consequently, the shell, contracting from infinity to the zero radius, should intersect the points of the sign changing of $\sigma_{\rm in}$ and $\sigma_{\rm out}$, which now should be placed in the corresponding $T-$regions. At the same time, it should be $\rho_{\sigma_{\rm out}} < r_g$ and $r_- < \rho_{\sigma_{\rm in}} < r_+$, which is evident from the graph in Fig. 55 for the behavior of r_g, $r_\pm$, ρ_0, $\rho_{\sigma_{\rm in}}$, $\rho_{\sigma_{\rm out}}$ and ρ_0 in dependance of $m_{\rm out}$.

The Carter–Penrose diagram for infinite motion of the self-attractive shell in the case $m_{\rm out} > m_{\rm out,1}$ is shown in Fig. 56 and coincides with similar one in in Fig. 46.

Finite Motion at $\Delta m > -M$ and $e^2 > GM^2$

It is remained to consider the finite motion at $\Delta m > -M$ and $e^2 > GM^2$. We already know, that in this case there is only one turning point

$$\rho_0 = \frac{B+\sqrt{D}}{2M\left(1-\frac{\Delta m^2}{M^2}\right)}, \qquad \frac{\partial \rho_0}{\partial(2Gm_{\rm out})} = \frac{\rho_0}{\sqrt{D}} > 0, \quad D \geq B^2 \geq 0. \tag{232}$$

Now we look on the inequalities between parameters e, M, $m_{\rm in}$ and $m_{\rm out}$:

$$Gm_{\rm in}^2 > e^2 > GM^2 \Rightarrow m_{\rm in} > M > -\Delta m, \tag{233}$$

$$\Delta m = m_{\rm out} - m_{\rm in} > -M \Rightarrow m_{\rm in} > m_{\rm out} > m_{\rm in} - M > 0. \tag{234}$$

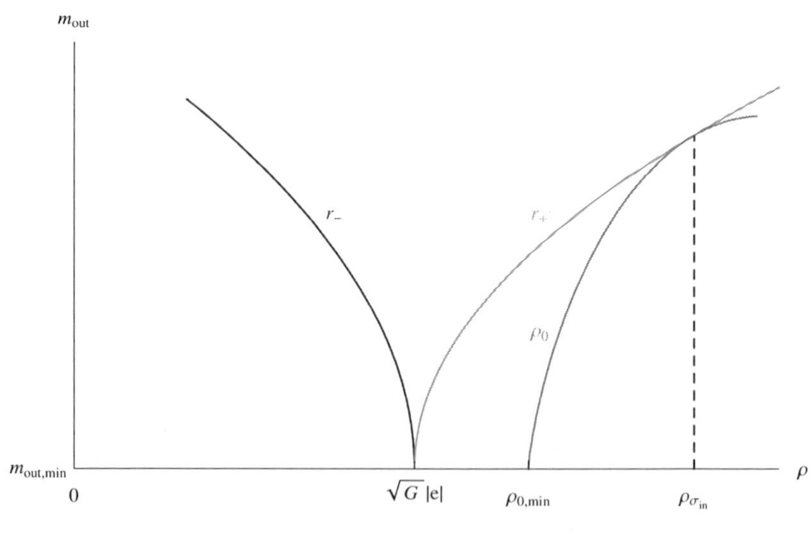

Figure 57. The relative positions of the characteristic radii ρ_0, $e\sqrt{G} < \rho_{0,\rm min} < \rho_{\sigma_{\rm in}}$ and $r_\pm$.

This means, that we cannot start our consideration from $m_{\rm out} = 0$ as previously. Now we need to to start from the nonzero minimal value $m_{\rm out,min} > 0$ from (209). This minimal value $m_{\rm out,min}$ in (209) at the fixed bare mass M depends on the corresponding minimal value of $m_{\rm in,min}$ at the fixed charge e. At the same time, the expression for $\rho_{0,\rm min}$ coincides with (214). It is easily to verify, that relation $\sqrt{D_{\rm min}} = e^2 + GM^2 + 2\sqrt{G}\Delta m e > 0$ in (213) is equivalent to inequalities $e\sqrt{G} < \rho_{0,\rm min} < \rho_{\sigma_{\rm in}}$. Graphically these inequalities are shown in Fig. 57.

From the point of view of the internal metric at $m_{\rm out,min} < m_{\rm out} < m_{\rm cr} + \Delta m$, where $m_{\rm cr}$ is defined by expression (215), the shell, starting from the zero radius expands up to the turning point, by passing through the T_+–region between the both horizons without changing the sign of $\sigma_{\rm in}(\rho_0) = +1$. Therefore, ρ_0 is placed in the (right) R_+–region beyond the horizon. At the same time, if $m_{\rm out} > m_{\rm cr} + \Delta m$, then the turning point is placed beyond the horizon r_+, but in the (left) R_-–region.

How all this looking from the point of view of the external metric? Now it is needed to compare the relative values of $\rho_{0,\rm min}$, $\rho_{\sigma_{\rm out}}$ and $r_{g,\rm min}$. It appears, that always $r_{g,\rm min} \leq \rho_{0,\rm min}$:

$$r_{g,\rm min} = 2\sqrt{G}e + 2G\Delta m \quad < \quad \rho_{0,\rm min} = \frac{GM^2 - e^2}{2(M - m_{\rm in})} \quad \Rightarrow \tag{235}$$

$$(e - \sqrt{G}M + 2\sqrt{G}\Delta m)^2 \quad \geq \quad 0. \tag{236}$$

If the condition $e - \sqrt{G}M + 2\sqrt{G}\Delta m < 0$ is held in (236), it is verified by direct substitution, that $\rho_{\sigma_{\rm out}} < \rho_{0,\rm min}$. The last relation provides also the inequality $\rho_{\sigma_{\rm out}} < r_{g,\rm min}$:

$$\frac{e^2 - GM^2}{-2\Delta m} < 2G\left(\frac{e}{\sqrt{G}} + \Delta m\right) \Rightarrow \tag{237}$$

$$(e - \sqrt{G}M + 2\sqrt{G}\Delta m)(e + \sqrt{G}M + 2\sqrt{G}\Delta m) < 0. \tag{238}$$

As we already know, the function in the second parenthesis is always positive under the considered conditions. This means, that the shell, starting from the zero radius, passes through the T_+–region of the external metric, the sign of $\sigma_{\rm out}$ is changing, and come out to the R_-–region, where $\sigma_{\rm out}(\rho_0) = -1$.

The corresponding Carter–Penrose diagrams for finite motion of the self-gravitating shell in the case $\Delta m < -M$ and $e^2 > GM^2$, $e - \sqrt{G}M + 2\sqrt{G}\Delta m < 0$ and $m_{\rm in} \gtrless m_{\rm cr}$ are shown in Figs. 58.

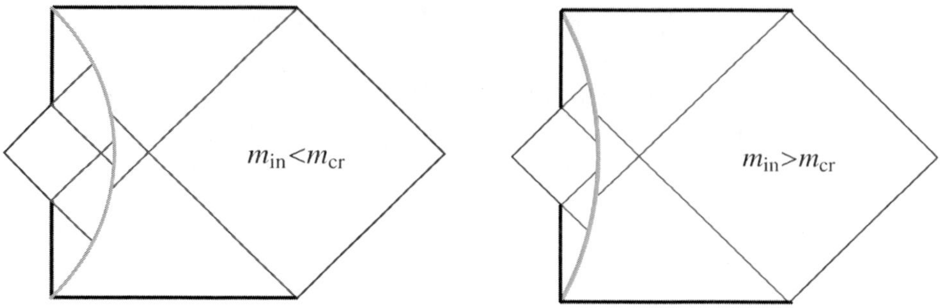

Figure 58. The Carter–Penrose diagrams for finite motion of the self-gravitating shell in the case $\Delta m < -M$ and $e^2 > GM^2$, $e - \sqrt{G}M + 2\sqrt{G}\Delta m < 0$ and $m_{\rm in} \gtrless m_{\rm cr}$.

Now, at last, we consider the case of $e - \sqrt{G}M + 2\sqrt{G}\Delta m > 0$, when $r_g < \rho_{0,\rm min} < \rho_{\sigma_{\rm out}} < \rho_{\sigma_{\rm in}}$. Graphically these inequalities are shown in Fig. 59.

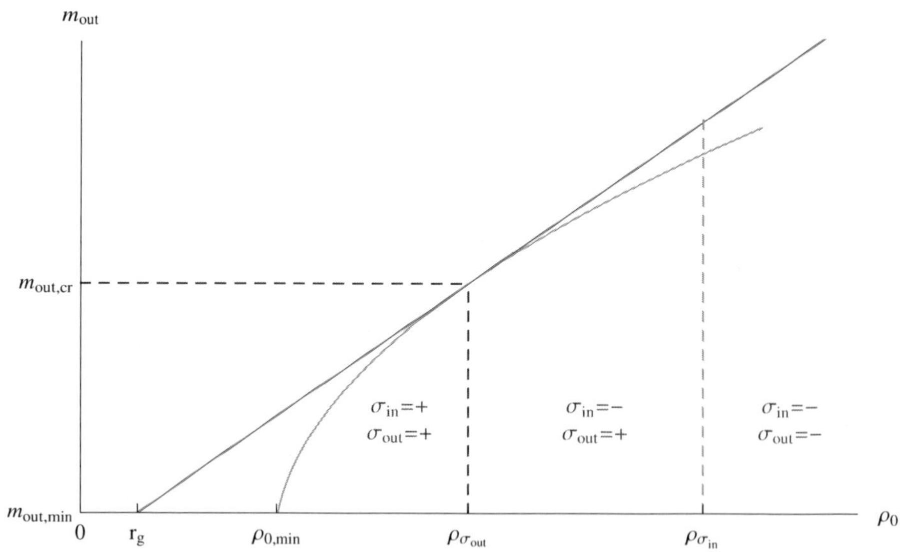

Figure 59. The relative positions of the characteristic radii r_g, $\rho_{0,\rm min}$, $\rho_{\sigma_{\rm in}}$ and $\rho_{\sigma_{\rm out}}$.

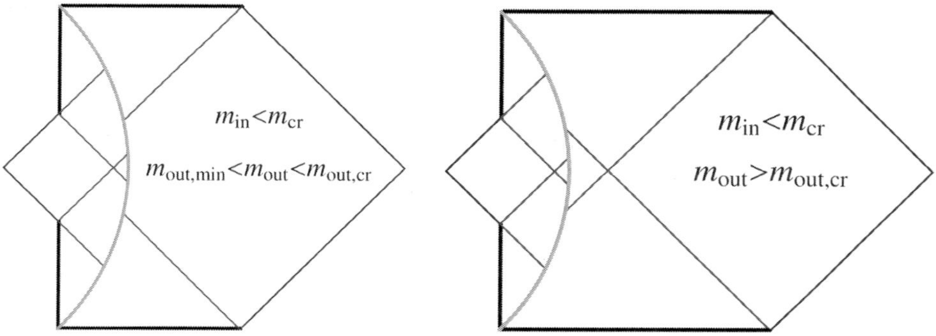

Figure 60. The Carter–Penrose diagrams for finite motion of the self-gravitating shell in the case $m_{in} < m_{in,cr}$ and $m_{out,min} < m_{out} < m_{out,cr}$ (the left panel), and, respectively, in the case of $m_{in} < m_{in,cr}$, $m_{out} > m_{out,cr}$ (the right panel).

We see, that there are three regions with the different combinations of σ_{in} and σ_{out}. In the first case, at $m_{in} < m_{in,cr}$ and $m_{out,min} < m_{out} < m_{out,cr}$, where

$$m_{out,cr} = -\frac{e^2 - GM^2}{2G\Delta m}, \tag{239}$$

there are realized $\sigma_{in}(\rho_0) = +1$ and $\sigma_{out}(\rho_0) = +1$. The corresponding Carter–Penrose diagram is shown at the left panel in Fig. 60.

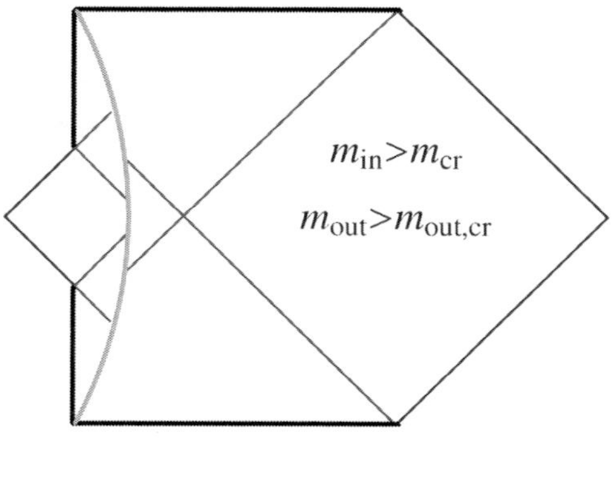

Figure 61. The Carter–Penrose diagrams for finite motion of the self-gravitating shell in the case $m_{in} > m_{in,cr}$, $m_{out} > m_{out,cr}$.

In the second case, at $m_{\rm in} < m_{\rm in,cr}$, but $m_{\rm out} > m_{\rm out,cr}$, there are realized $\sigma_{\rm in}(\rho_0) = +1$ and $\sigma_{\rm out}(\rho_0) = -1$. This second case corresponds the the left panel in Fig. 60.

At last, in the third case, at $m_{\rm in} > m_{\rm in,cr}$ and $m_{\rm out} > m_{\rm out,cr}$, it is realized $\sigma_{\rm in}(\rho_0) = -1$ and $\sigma_{\rm out}(\rho_0) = -1$. The corresponding Carter–Penrose diagram for this third case is shown in Fig. 61.

In result we obtain the complete classification of the possible types of the spherically symmetric global geometries for two types of electrically charged shells:

(1) The charged shell as a single source of the gravitational field, when internal space-time is flat, and external space-time is the Reissner–Nordström metric;

(2) The neutralizing shell with an electric charge opposite to the charge of the internal source with the Reissner–Nordström metric and with the Schwarzschild metric outside the shell.

Chapter 7

Reissner–Nordström Metric with Neutral Shell

The complexity of the Einstein equations dictates consideration of very simple and highly symmetric models. Apart from the well known examples of cosmological models, the general relativistic effects and therefore, the most interesting deviations from Newtonian gravitation should be expected when considering the very concentrated massive bodies. The simplest of them, of course, is the point-like particle. The space-time around the point-like particle appeared to be the first exact solution to the Einstein vacuum equations found by Karl Schwarzschild in 1916, just few months after presentation of the new gravitational theory in the Berliner Academy of Sciences by Albert Einstein. This solution was spherically symmetric and looked, at the first sight, as a simple generalization of the Newton's gravity law.

Only in 1960's it became clear that the Schwarzschild space-time has rather nontrivial global geometry: it possesses an event horizon (which serves as a black hole boundary), two asymptotically flat regions and two singularities at zero radii, where the gravitating sources are concentrated. The appearance of event horizons separating the space-time region from where the light rays can escape to spatial infinity and those from where it is impossible to do this, is just the consequence of Special Relativity, namely, it follows from the fact that photons can not be stopped, and this exhibits the inconsistency between Special Relativity and Newtonian gravitation. In the Schwarzschild space-times there are two branches of the horizon (the past and the future ones called the particle and event horizons, respectively), and beyond them there lie two unusual regions: of inevitable expansion (in the past) and inevitable contraction (in the future. It is in such regions where the singularities appeared. These singularities are also unusual, they are space-like. Thus, the source in the Schwarzschild space-time exists only one moment in at zero radius in the past, then disappears and resurrects in the future singularity, again at zero radius. Thus, such a source is clearly nonphysical and can not be considered (even theoretically) as a limit of any extended static body.

In a more general case, outside the point-like electrically charged massive source, the solution to the electro-vacuum Einstein equations was found by Reissner and Nordström. The global geometry of a Reissner–Nordström manifold depends on the relation between the electrical charge and the total mass (energy) of the system (the latter includes both the energy of Coulomb field and the binding gravitational energy). For small enough charge the manifold (which is called in this case the Reissner–Nordström black hole) consists of the same elements as the Schwarzschild ones. But the number of these parts is infinite and they form an infinitely long ladder from the past to the future. In addition to the event (and particle) horizons, called now the outer horizons (infinitely many of them), there exist the inner horizons that serve as Cauchy horizons beyond which the trajectories of test particles, whose initial data are specified in the one of the asymptotically flat regions, cannot be unambiguously continued. What concerns the singularities at zero radii with sources (of course, there are also infinitely many of them), they are time-like, but hidden beyond the horizons of both types and, thus, can not be thought of as the limits of any usual extended static charged body. When the charge grows while the total mass remains constant, the outer and inner horizons become closer and closer to each other and they merge eventually for some critical value of charge (which is proportional to the total mass). Such a manifold with the double horizon is called the extreme Reissner–Nordström black hole. The singularity with the source is still hidden beyond the horizon in this case, Finally, if the charge exceeds the critical value, the horizons disappear, and the naked singularity reveals itself. In this case the global geometry is no more an infinite ladder, but it rather resembles that of the flat Minkowskian space-time with only one, but important difference: the zero radius time-like world-line is now singular and contains the charged massive source. But, physically, it is not a limit of any extended static charged body because the latter is unstable and would expand infinitely without some additional force (which, in turn, would add some energy-mass to the whole system).

The simplest generalization of the point-like particle is a spherically symmetric thin neutral dust shell. Though it is also singular (the finite amount of mass=energy is concentrated in an infinitesimal volume), but the singularity is now spread on a sphere of finite radius, there are both interior and exterior regions where the space-time metrics are, in principle, known. These are parts of Schwarzschild and Reissner–Nordström manifolds, whose parameters are related to that ones describing the immersed shell and its initial state. The evolution of thin shell is governed by the so called Israel equations. The latter are nothing but the matching conditions between the inner and outer metrics. The mathematical structure of these equations reflects that of Einstein equations (of course, in three dimensions instead of four): there are both constraints and dynamical equations. Due to the spherical symmetry we will have only one constraint and one dynamical radial equation plus their differential consequence — analog of Bianchi identity, which is nothing more but the continuity equation relating the surface energy density of the shell and its surface tension. In principle, given the shell's equation of state, ir is possible to solve this continuity equation and, thus we are left with only one equation, the constraint. And it is this equation that will be the subject of our investigation. The final goal is the construction of the so called Carter–Penrose conformal diagrams that describe quite clearly the global geometry of the composite manifold for every combination of the parameters involved.

We have already mentioned that the global geometry of the vacuum and electro-vacuum spherically symmetric solutions to the Einstein equations is by no means trivial. For the composite manifolds with the shells and different solutions inside and outside it, the number of different combinations becomes very large. To simplify their investigation we need some method in order to recognize easily which of them is realized for any particular choice of the parameters of, say, an inner metrics and a shell. In this paper we propose the effective potential method. Drawing together the effective potential an two more very simple curves we are able to construct immediately the Carter–Penrose diagram for any allowed values of the shell's parameters.

The paper is organized as follows. In the Section "Preliminaries" we describe shortly the spherical gravity, the construction of Carter–Penrose conformal diagrams, the thin shell formalism and the method of the effective potential. The subsequent Sections are devoted to the application of the proposed method to the construction of the global geometries for neutral spherically symmetric thin dust shells immersed into different Reissner–Nordström manifolds.

Throughout the paper we use units with $\hbar = c = 1$, where $\hbar$ is the Planckian constant, and c is the speed of light.

Chapter 8

Preliminaries

8.1. Spherical Gravity

The structure of any spherically symmetric space-time is completely determined by two invariant functions of two variables. Indeed, locally, the general spherically symmetric metric can be written as

$$ds^2 = A^2 dt^2 + 2H dt dq - B^2 dq^2 - R^2 d\sigma^2 \,, \tag{240}$$

where $A(t,q), H(t,q)$ and $B(t,q)$ are functions of the time coordinate, t, and some radial coordinate, q, $d\sigma^2$ is the line element of a $2-dim$ unit sphere, and $R(t,q)$ is the radius of this sphere in the sense that its area equals $4\pi R^2$. Therefore, we are, actually, dealing with the invariant function $R(t,q)$ and the two-dimensional metric, which by suitable coordinate transformation can always be put in the conformally flat form

$$ds_2^2 = \gamma_{ik} dx^i dx^k = \omega^2(t,q)(dt^2 - dq^2) \,, \quad i,k = 0,1 \,. \tag{241}$$

This proves the above statement about two functions of two variables.

The first invariant function is, of course, the radius $R(t,q)$. By geometrical reasons, we choose for the second function the invariant (notations are obvious)

$$\Delta = \gamma^{ik} \frac{\partial R}{\partial x^i} \frac{\partial R}{\partial x^k} = \frac{1}{\omega^2} \left(\dot{R}^2 - R'^2\right) . \tag{242}$$

This is nothing more but the square of the normal vector to the surfaces of constant radii, $R(t,q) = const$. The invariant function Δ brings a very important geometrical information. If $\Delta < 0$, the surfaces $R = const$ are time-like, such regions are called the $R_\pm$–regions, the signs "$\pm$" being denote the sign of a spatial derivative of the radial function R. If $\Delta > 0$, the regions are called the $T_\pm$–regions, depending on the sign of the corresponding time derivative (inevitable expansion or inevitable contraction), and the surfaces $R = const$ are space-like. The $R_\pm-$ and $T_\pm-$ regions are separated by the apparent horizons with $\Delta = 0$. It is the set of these regions and horizons together with the boundaries (infinities and that determines the global geometry. The boundaries are to be chosen in such a way that the space-time becomes geodesically complete, namely, all the time-like and null geodesics should start and end either at infinities or at singularities.

8.2. Geodesically Complete Conformal Carter–Penrose Diagrams

The causal structure of geodesically complete spherically symmetric space-times can be best seen on the conformal Carter–Penrose diagrams where each point represents a sphere, and infinities are brought to the final distances. Since every 2-dimensional space-time is (locally) conformally flat, its Carter–Penrose diagram is the set of that for the 2-dimensional Minkowski manifold. To see how the latter looks like, let us, at first, transform the Minkowski metric $ds^2 = dt^2 - dx^2$ to the double-null coordinates $u = t - x$ (retarded time) and $v = t + x$ (advanced time), then $ds^2 = du\,dv$. We will use the convention that on the diagram the time coordinate increases from down to up, the spatial coordinate — from left to right, and the null curves $u = const$, $v = const$ are the straight lines with the slope $\pm 45°$. Making one more transformation

$$\begin{aligned} u' &= \arctan u\,, \quad -\frac{\pi}{2} \le u' \le \frac{\pi}{2} \\ v' &= \arctan v\,, \quad -\frac{\pi}{2} \le v' \le \frac{\pi}{2} \end{aligned} \tag{243}$$

one gets

$$\begin{aligned} ds^2 &= \Omega^2 ds'^2\,, \quad \Omega = \frac{1}{\cos u' \cos v'} \\ ds'^2 &= du'dv' = dt'^2 - dx'^2\,. \end{aligned} \tag{244}$$

Formally, the metric ds'^2 looks exactly as the starting one, but now coordinates $(u',\, v')$ and $(t',\, x')$ run the finite intervals.

The Carter–Penrose diagram for the complete 2-dimensional Minkowski space-time $(-\infty < t < \infty,\ -\infty < x < \infty)$ is shown in Fig. 62. Here $J^{\pm}(J'^{\pm})$

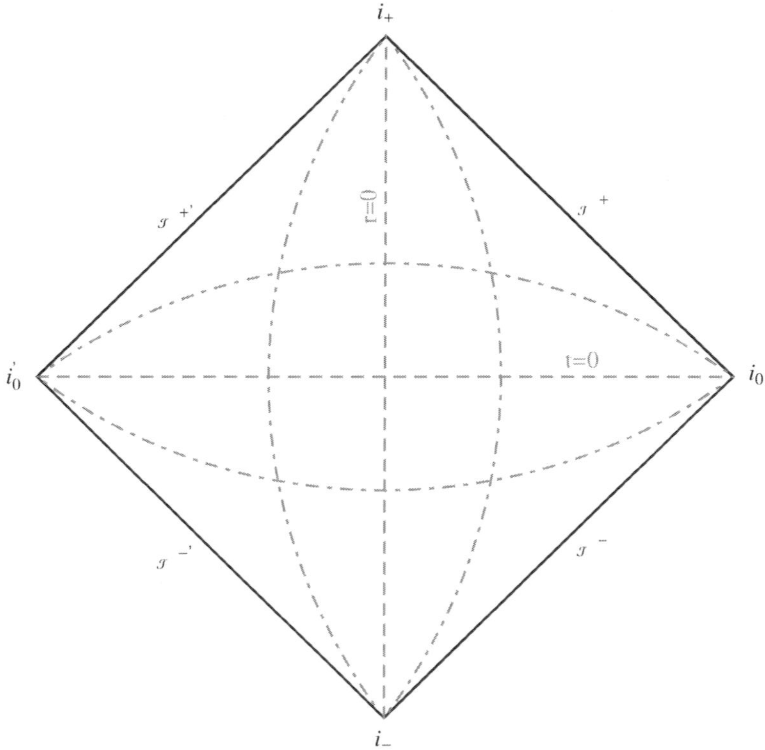

Figure 62. The Carter–Penrose diagram for the complete 2-dimensional Minkowski space-time $(-\infty < t < \infty,\ -\infty < x < \infty)$. The horizontal dashed curves represent $t = const$ lines, while the vertical ones are for $x = const$.

are null future $(v'(u') = \pi/2, v(u) = \infty)$ and past $(u'(v') = -\pi/2, u(v) = -\infty)$ infinities, $i_{\pm}$ are future and past $(t' = \pm\pi/2)$ temporal infinities, and $i_0(i_0')$ are spatial $(x' = \pm\pi/2, x = \pm\infty)$ infinities. If the corresponding conformally flat metric is not complete in the sense that one of the coordinates starts from or ends at the finite boundary value (like, for example, the zero radius value in the case of spherical symmetry), then one should cut the above square along the corresponding diagonal (in general, along some time-like os space-like curve), and such part of the complete Carter–Penrose diagram will be a triangle with the vertical (left for R_+–regions and right for R_-–regions) or horizontal (for $T_\pm$–regions) boundary.

Both the Schwarzschild and Reissner–Nordsrtöm metrics look the same in the so-called curvature coordinates:

$$ds^2 = F dt^2 - \frac{1}{F} dR^2 - R^2 (d\vartheta^2 + \sin^2\vartheta d\varphi^2), \tag{245}$$

where R — radius ($0 \leq R < \infty$), $F = F(R)$, and ϑ and φ are spherical angles. The two-dimensional part can easily be written in the conformally flat form by introducing the "tortoise" coordinate $R^\star$:

$$\begin{aligned} dR^\star &= \frac{dR}{|F|}, \\ ds_2^2 &= F\left(d\xi^2 - dR^{\star 2}\right). \end{aligned} \tag{246}$$

In the $R_\pm$–regions $F = -\Delta > 0$ and $R^\star$ plays the role of the spatial (radial) coordinate q, while ξ is the time coordinate t. In the $T_\pm$–regions, $R^\star$ plays the role of the time coordinate t, while ξ is the spatial coordinate q.

Consider, first, the Schwarzschild metric. In this case

$$F = 1 - \frac{2Gm}{R}, \tag{247}$$

where G is the Newton's gravitational constant m is the total mass of the gravitating system measured by distant observers (at infinity), and we put the speed of light $c = 1$. For $R > r_g = 2Gm$ we have the the R–region, and for $R < r_g$ — the T–region. The event horizon coincides with the apparent horizon at $R = r_g$ (gravitational, or Schwarzschild, radius). At $R = 0$ we encounter the (space-like) curvature singularity. The complete Carter–Penrose diagram looks as follows in Fig. 63. There are two isometric $R_\pm$–regions bounded by two apparent (past and future) horizons at $R = r_g$ and two asymptotically flat regions with corresponding future and past temporal ($i_\pm, i'_\pm$), future and past null ($J_\pm, J'_\pm$) and spatial (i_0, i'_0) infinities. Also we have two T–regions (T_+ and T_-) bounded by the apparent horizons at $R = r_g$ and future and past space-like singularities at $R = 0$. This is called the eternal Schwarzschild black hole. The gravitational source is concentrated on these two space-like singularities, i.e., it exists only for one moment in the past and reappears again for one moment in the future.

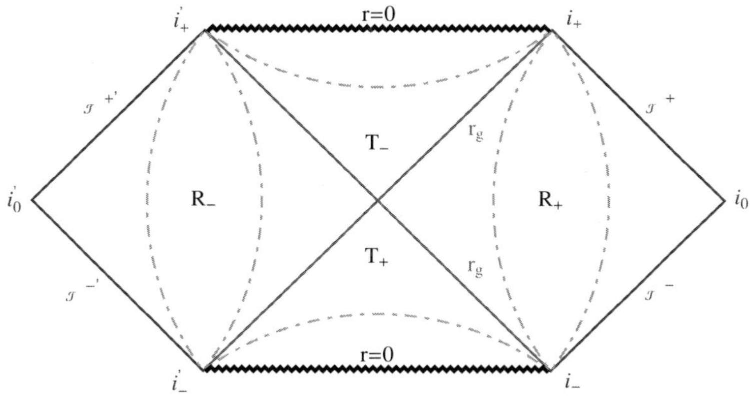

Figure 63. The complete Carter–Penrose diagram of the Schwarzschild metric.

The causal structure of the Reissner–Nordström space-time is much more complex. The function F equals now

$$F = 1 - \frac{2Gm}{R} + \frac{Ge^2}{R^2}, \tag{248}$$

e is the electric charge. There are three different cases

(1) $Gm^2 > e^2$ — Reissner–Nordström black hole, equation $F = 0$ has two nonequal real roots $r_{\pm}$,

$$r_{\pm} = Gm \pm \sqrt{G^2m^2 - Ge^2}. \tag{249}$$

According to the signs of F, we have the R–regions for $r_+ < R < \infty$ and $0 \le R < r_-$, T–regions in-between, $r_- < R < r_+$, and two apparent horizons at $R = r_{\pm}$, the external one, r_+, playing the role of the event horizon, and the inner, r_-, — the Cauchy horizon. The geodesically complete Carter–Penrose diagram is the ladder extended infinitely to the past and to the future as is shown in Fig. 64.

In the complete (eternal) Reissner–Nordström black hole space-time both the the gravitational source and the electric charge(s) are concentrated on two (for each part of the ladder) time-like singularities $R = 0$ (left and right on the diagram), the signs of the electric charges on them being opposite..

(2) $Gm^2 = e^2$ — extreme Reissner–Nordström black hole. Equation $F = 0$ has the double root $r_+ = r_- = Gm = \sqrt{G}|e|$. We have R–regions everywhere except the apparent (event) horizon at $R = r_+ = r_-$, as is shown in Fig. 65.

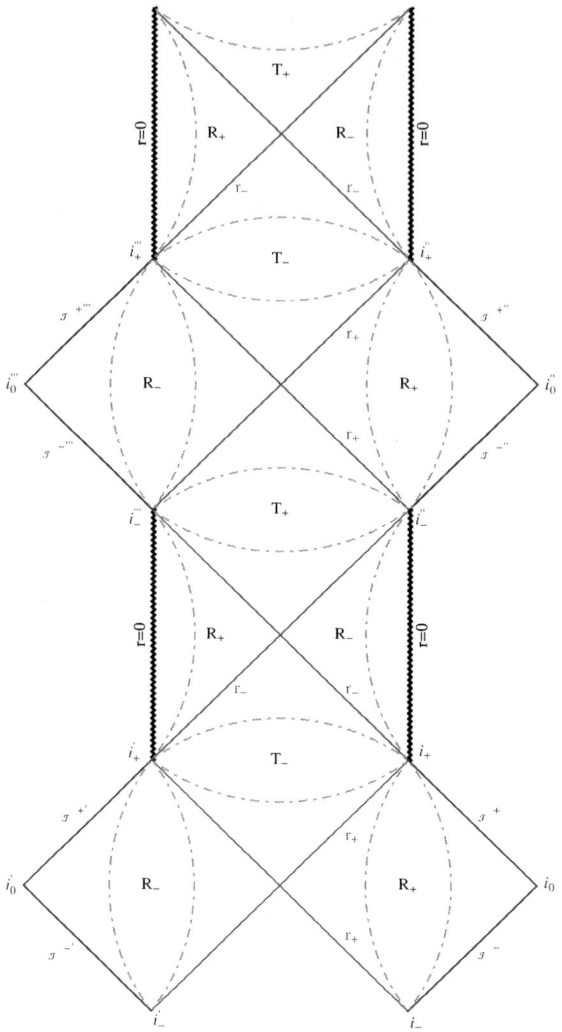

Figure 64. The complete Carter–Penrose diagram of the Reissner–Nordström black hole, $Gm^2 > e^2$.

(3) $Gm^2 < e^2$ — no black hole, the naked singularity at $R = 0$. The Carter–Penrose diagram is very simple (see Fig. 66).

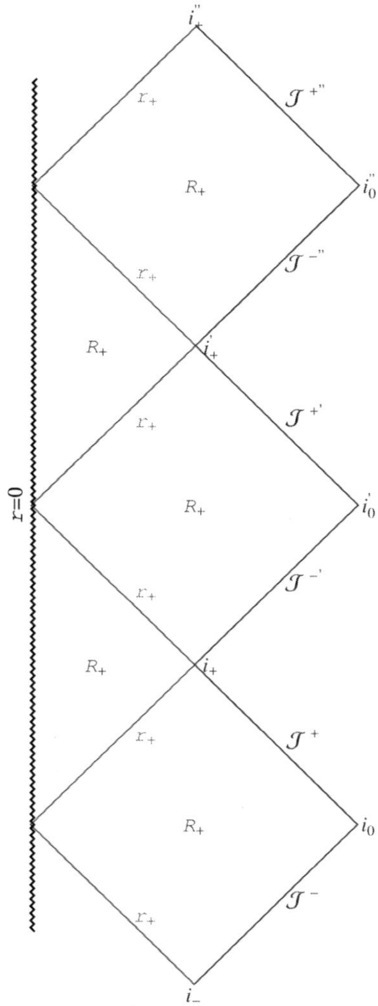

Figure 65. Extremal Reissner–Nordström black hole, $Gm^2 = e^2$.

8.3. Thin Shells

The thin shell is a hyper-surface in the space-time on which the energy-momentum tensor is singular. If such a hyper-surface is time- or space-like, one can introduce in its vicinity the so-called Gaussian normal coordinates, and the line

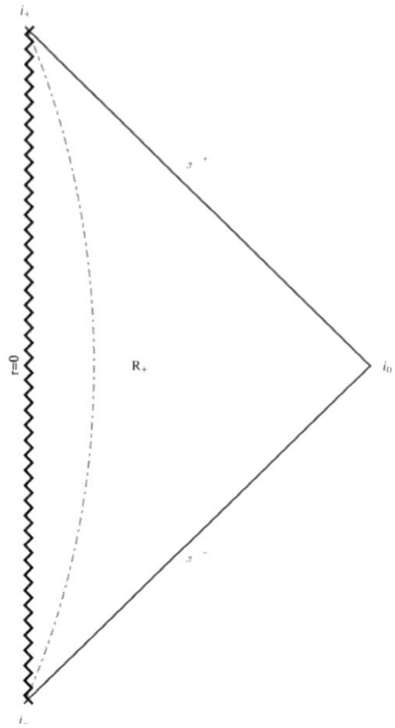

Figure 66. The complete Carter–Penrose diagram of the Reissner–Nordström naked singularity, $Gm^2 < e^2$.

element can be written as

$$ds^2 = \varepsilon dn^2 + \gamma_{ij}(n,x)dx^i dx^j\,, \tag{250}$$

n is the coordinate in the normal direction to the shell, and x^i — coordinates on the shell, $\varepsilon = +1$ in the space-like case and $\varepsilon = -1$ in the time-like case. The surface is supposed to be located at $n = 0$. The energy-momentum tensor T^μ_ν is proportional to δ-function,

$$T^\mu_\nu = S^\mu_\nu \delta(n)\,, \tag{251}$$

S^μ_ν is called the surface energy-momentum tensor. The dynamics of the thin shell is governed by the Israel equations obtained by integrating the Einstein equations across the shell. First of all, one gets $S^n_n = S^i_n = 0$, this can be considered as the definition of the thin shell. The Israel equations are

$$\varepsilon\left([K_{ij}] - \gamma_{ij}\,[K]\right) = 8\pi G S_{ij}\,, \tag{252}$$

supplemented by the Bianchi identity for the shell

$$S^j_{i|j} + [T^n_i] = 0. \tag{253}$$

Here $K_{ij} = -(1/2)\partial\gamma_{ij}/\partial n$ is the extrinsic curvature tensor, K is its trace, brackets $[\,] = (out) - (in)$ is the jump across the shell, the vertical line denotes the covariant derivative with respect to the metric γ_{ij}. In what follows we will be dealing with the time-like shells only, so, $\varepsilon = -1$.

In the case of spherical symmetry everything is simplified drastically. The metric becomes

$$ds^2 = -dn^2 + \gamma_{00}(n,\tau)d\tau^2 - \rho^2(n,\tau)d\sigma^2, \tag{254}$$

$\rho(0,\tau)$ is the shell radius as a function of the proper time of the observer sitting on this shell, $n < 0$ inside and $n > 0$ outside. The mixed components of the surface energy momentum tensors are S^0_0 (surface energy density) and $S^2_2 = S^3_3$ (surface tension), and the Israel equations reduced to one constraint and one dynamical equations, namely,

$$\begin{aligned} \left[K^2_2\right] &= 4\pi G S^0_0 \\ \left[K^0_0\right] + \left[K^2_2\right] &= 8\pi G S^2_2 . \end{aligned} \tag{255}$$

The supplement equation is now

$$\dot{S}^0_0 + \frac{2\dot{\rho}}{\rho}\left(S^0_0 - S^2_2\right) + [T^n_0] = 0. \tag{256}$$

We are interested in the situation when both inside and outside the shell the space-time is (electro)-vacuum one, hence, $T^n_0 = 0$. For the sake of simplicity we will consider the dust shell, for which $S^2_2 = 0$. Then,

$$S^0_0 = \frac{M}{4\pi\rho^2}, \tag{257}$$

where $M = const$ is the bare mass of the shell (without the gravitational mass defect). Thus, we need only the first, constraint, equation. In order to go further we have to calculate $K^2_2 = -(1/\rho^2)K_{22} = -1/(2\rho^2)\partial\rho^2/\partial n = -\rho_{,n}/\rho$. But, from definition of the invariant Δ it follows

$$\begin{aligned} \Delta &= \dot{\rho}^2 - \rho^2_{,n} \\ \rho_{,n} &= \sigma\sqrt{\dot{\rho}^2 - \Delta} \\ K^2_2 &= -\frac{\sigma}{\rho}\sqrt{\dot{\rho}^2 - \Delta}. \end{aligned} \tag{258}$$

Here $\sigma = \pm 1$ depending on whether radii increasee ($\sigma = +1$) in the normal outward direction or decrease ($\sigma = -1$). Thus, the sign of σ coincides with that of the R–region, and it can change only in the T–regions. Finally, the only equation we will need in our analysis is

$$\sigma_{\text{in}}\sqrt{\dot{\rho}^2 - \Delta_{\text{in}}} - \sigma_{\text{out}}\sqrt{\dot{\rho}^2 - \Delta_{\text{out}}} = \frac{GM}{\rho}. \tag{259}$$

Since in our case $\Delta = -F$, we have

$$\sigma_{\text{in}}\sqrt{\dot{\rho}^2 + 1 - \frac{2Gm_{\text{in}}}{\rho} + \frac{Ge_{\text{in}}^2}{\rho^2}} - \sigma_{\text{out}}\sqrt{\dot{\rho}^2 + 1 - \frac{2Gm_{\text{out}}}{\rho} + \frac{Ge_{\text{out}}^2}{\rho^2}} = \frac{GM}{\rho}. \tag{260}$$

We will not consider exotic matter shells, so $M > 0$. From the above constraint equation (that is nothing more but the energy conservation law) it follows that for the qualitative analysis one needs to investigate the behavior of the function $\rho(\tau)$ only at several special points: $\rho \to \infty$, $\dot{\rho} = 0$, $\rho = 0$ and $\rho = \rho_\sigma$ where $\sigma_{\text{out}}(\sigma_{\text{in}})$ changes its sign.

Chapter 9

The Effective Potential Method

Based on the Israel equations for dynamics of thin shells, in this Section we introduce the so called effective potential which enables us to construct the conformal Carter–Penrose diagrams, describing clearly the global geometry of the corresponding composite space-time manifold, without thorough analytical investigations.

In what follows we will consider the spherically symmetric neutral thin dust shells immersed into different types of Reissner–Nordström space-times. So, both inside and outside the shell of bare mass M we have the same value of electric charge e but different mass parameters m_{in} and m_{out}. We already discussed in the preceding Section that of all Israel equations the only one we need in our case is the following constraint equation

$$\sigma_{\text{in}}\sqrt{\dot{\rho}^2+1-\frac{2Gm_{\text{in}}}{\rho}+\frac{Ge^2}{\rho^2}}-\sigma_{\text{out}}\sqrt{\dot{\rho}^2+1-\frac{2Gm_{\text{out}}}{\rho}+\frac{Ge^2}{\rho^2}}=\frac{GM}{\rho}. \quad (261)$$

Here ρ is the shell's radius as a function of the proper time τ (the over dot means its first derivative). Let us remind that in chosen units its square root has dimension of length, time and inverse mass simultaneously. The sign functions $\sigma_{in,out}=\pm1$ in the interior and exterior parts of the complete manifold show us whether the radii r increase (+) or decrease (-) in the outward normal direction to the shell. It is the values of $\sigma's$ that define, essentially, the global geometry of the composite manifold. To specify the solution to this differential equation we need the initial data. In addition to the usual initial values of the shell's radius ρ_0 and rapidity $\dot{\rho}_0$ we should also know the initial value of σ_{in}. Given, then, the parameters of our system, namely, the mass m_{in} and the electric charge e of the inner metrics and the bare mass M of the shell itself, we are able to calculate both σ_{out} of the outer metrics and its mass parameter, m_{out}, which is the total mass (= energy) of the whole system (for this very reason the constraint equation is often called the equation of initial conditions). It is convenient to choose as initial data either the value of radius at (one of) the turning points where $\dot{\rho}_0 = 0$, whenever they exist, or the rapidity at infinity. Then the initial values of σ_{in} and σ_{out} show us, in which of the $R_{\pm}$–regions of the inner and outer parts of the complete manifold the shell starts to move. The signs of $\sigma' s$ may change their initial values dynamically, but only in $T_{\pm}$–regions (provided they exist) when the expressions under the corresponding square roots become equal zero.

The experience in classical dynamics show us that the most convenient method to visualize particle motion under influence of different forces is the construction (whenever it is possible) the corresponding potentials. In such cases the first integral of dynamical equations (i.e. those containing the second time derivatives — accelerations) equals the total energy of a particle which is just the sum of the kinetic energy and the potential energy. When the Newtonian (non-relativistic) gravitational force is acting, the potential energy is equal to the inertial (= gravitational) mass of the particle times the so called gravitational potential. Drawing together the potential energy graph and the horizontal line, corresponding to the particle energy, one can obtain very useful qualitative information. Namely, the allowed interval for the particle motion is that where the energy line lies above the potential curve, the minima of the potential being correspond to the stable equilibrium, while its maxima — to the unstable ones. Also it is easy to see when the motion is bound or unbound.

In our case of the spherical symmetric thin dust shell the situation is more tricky. First of all, though the constraint, (261), is actually the first integral of the corresponding dynamical equation (after solving the shell's continuity equation), it does not have the familiar "energy" form. The latter is recovered in the non-relativistic limit $\dot{\rho}^2 \ll 1$, the "particle" energy being identified with

$$\Delta m = m_{\text{out}} - m_{\text{in}}. \tag{262}$$

Fortunately, it is possible to put the constraint equation into (almost) such a form. Let us rewrite it in the following way

$$\sigma_{\text{in}}\sqrt{\dot{\rho}^2 + 1 - \frac{2Gm_{\text{in}}}{\rho} + \frac{Ge^2}{\rho^2}} - \frac{GM}{\rho} = \sigma_{\text{out}}\sqrt{\dot{\rho}^2 + 1 - \frac{2Gm_{\text{out}}}{\rho} + \frac{Ge^2}{\rho^2}}. \tag{263}$$

By squaring this relation one gets

$$\Delta m = \sigma_{\text{in}}\sqrt{\dot{\rho}^2 + 1 - \frac{2Gm_{\text{in}}}{\rho} + \frac{Ge^2}{\rho^2}} - \frac{GM^2}{2\rho} = \sigma_{\text{in}}\sqrt{\dot{\rho}^2 + 1 - \frac{2Gm_{\rho}}{\rho}} - \frac{GM^2}{2\rho}. \quad (264)$$

Here $m(\rho) = m_{\text{in}} - e^2/(2\rho)$ can be called "the running mass", because m_{in} is the total mass measured at spatial infinity of the complete inner space-time manifold, and $e^2/(2\rho)$ is the Coulomb energy outside the sphere of radius ρ, so, $m(\rho)$ is just the total mass (= energy) confined inside this sphere. The last term in (264) is the self-interaction energy of the shell. Though in our relativistic expression the kinetic and (gravitational) potential energies can not be separated, we may use it in the same way as in non-relativistic mechanics. Namely, we will consider $\Delta m(\dot{\rho} = 0)$ as the (now effective) potential. It is convenient to deal with dimensionless entities, so we adopt as definition the following

$$\frac{\Delta m}{M}(\dot{\rho} = 0) \equiv V_{\text{eff}} = \sigma_{\text{in}}\sqrt{1 - \frac{2Gm_{\text{in}}}{\rho} + \frac{Ge^2}{\rho^2}} - \frac{GM}{2\rho}. \quad (265)$$

Important note: the value of Δm may be both positive and negative (at least it is negative when $\sigma_{\text{in}} = -1$), i.e., the total mass (energy) of the whole system is less then that of the inner part of the space-time. But this should not confuse us. Let us remind that, by definition, the bare mass (energy) inside the spherical layer equals

$$M = 4\pi \int_{q_1}^{q_2} T_0^0 r^2(q) e^{\lambda/2} dq = 4\pi \int_{n_1}^{n_2} T_0^0 r^2(n)\, dn, \quad (266)$$

where the radial coordinate q (or the Gaussian normal coordinate n) runs from inside to outside (the shell is situated at $n = 0$), while the total mass (energy), which includes the gravitational mass defect is expressed by the Landau formula [2]

$$m = 4\pi \int_{r_1}^{r_2} T_0^0 r^2 dr = 4\pi \int_{n_1}^{n_2} T_0^0 r^2 r_{,n} dn = 4\pi \int_{n_1}^{n_2} T_0^0 r^2 \sigma |r_n|\, dn, \quad (267)$$

and for $\sigma_{\text{in}} = -1$ it is negative. One more thing: of course, we can get rid of the square root and obtain the expression more like the non-relativistic one, but, in doing this, we are loosing the very important information about the sign of σ_{in}, which is responsible for the global geometry of the inner part of the composite manifold. Surely, we already lost an information about σ_{out} due to first squaring, but it can easily be restored. Indeed, knowing σ_{in}, one obtains from Eqs. (261) and (267) that

$$\sigma_{out} = sign\left(\frac{\Delta m}{M} - \frac{GM}{2\rho}\right) \tag{268}$$

$$\sigma_{in} = sign\left(\frac{\Delta m}{M} + \frac{GM}{2\rho}\right). \tag{269}$$

We, thus, see that in order to reconstruct the global geometry of the composite space-time manifold and visualize qualitatively the thin shell motion, one needs to know not only the potential curve, $y_1 = V_{eff}(\rho)$, but, in addition, two more curves showing the change of signs σ_{in} and σ_{out}, namely, $y_2 = GM/(2\rho)$ and $\tilde{y}_2 = -GM/(2\rho) = -y_2$. In our Figures we will paint the effective potential curve y_1 (in red color), the curves y_2 and $\tilde{y}_2$ (in blue), and the $\Delta m/M = \mu = const$ — in horizontal (green) lines. The rules are the following:

(1) the regions allowed for shells motion are:

$$\mu > V_{eff} \quad \text{if} \quad \sigma_{in} = +1,$$
$$\mu < V_{eff} \quad \text{if} \quad \sigma_{in} = -1.$$

(2) how to determine the sign of σ_{in}:

$$\sigma_{in} = +1 \quad \text{if} \quad \mu > \tilde{y}_2,$$
$$\sigma_{in} = -1 \quad \text{if} \quad \mu < \tilde{y}_2.$$

(3) how to determine the sign of σ_{out}:

$$\sigma_{out} = +1 \quad \text{if} \quad \mu > y_2,$$
$$\sigma_{out} = -1 \quad \text{if} \quad \mu < y_2.$$

In the remaining part of the Section we will investigate the forms of the effective potential for different types of the inner Reissner–Nordström metrics, that depend on the values of the inner mass m and electric charge e, and different values of the shell's bare mass M. As usual, it is convenient to use some dimensionless variables and parameters. So, we fix the inner total mass m_{in} and introduce the dimensionless radius $x = \rho/(Gm_{in})$ and three dimensionless parameters, $\varepsilon = |e|/(\sqrt{G}m_{in})$, $\alpha = M/m_{in}$ and $\mu = \Delta m/M$. Then,

$$y_1 = \mu(\dot{\rho} = 0) \equiv V_{eff} = \sigma_{in}\sqrt{1 - \frac{2}{x} + \frac{\varepsilon^2}{x^2}} - \frac{\alpha}{2x}, \qquad y_2 = \frac{\alpha}{2x}, \qquad \tilde{y}_2 = -\frac{\alpha}{2x} \tag{270}$$

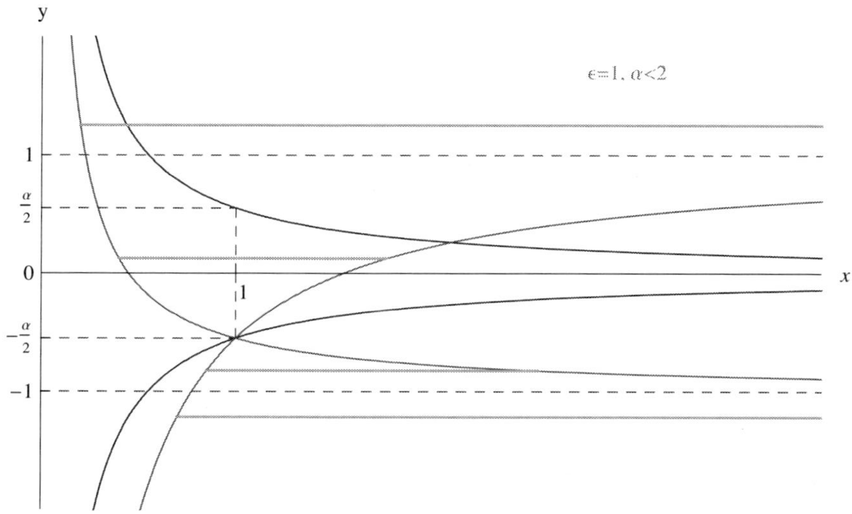

Figure 67. The effective potential for "light" shells.

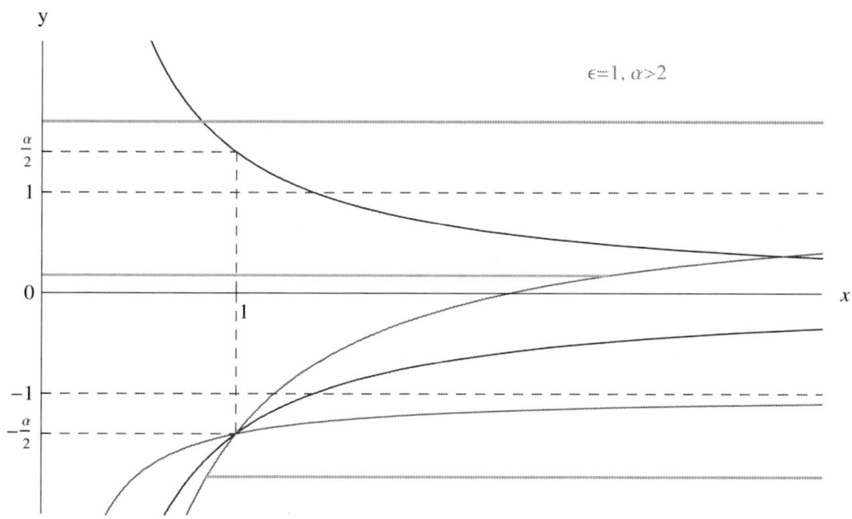

Figure 68. The effective potential for "heavy" shells.

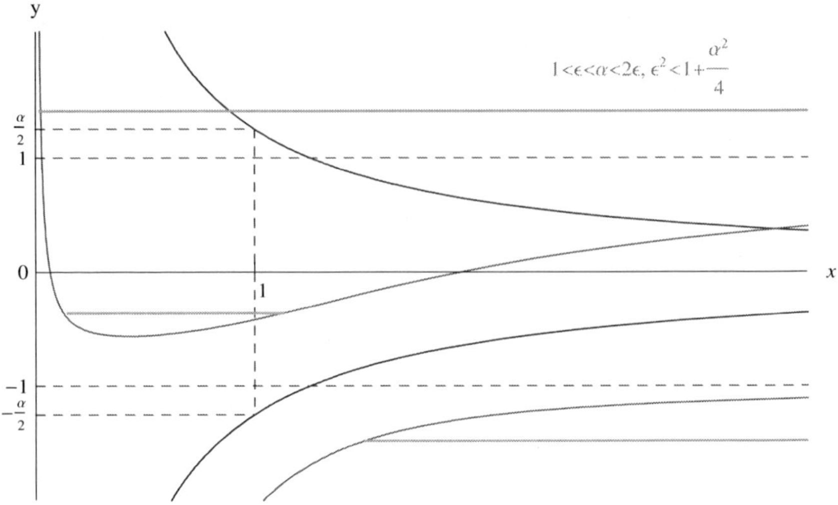

Figure 69. The effective potential at $1 < \varepsilon < \alpha < 2\varepsilon$ and $\varepsilon^2 < 1+\alpha^2/4$.

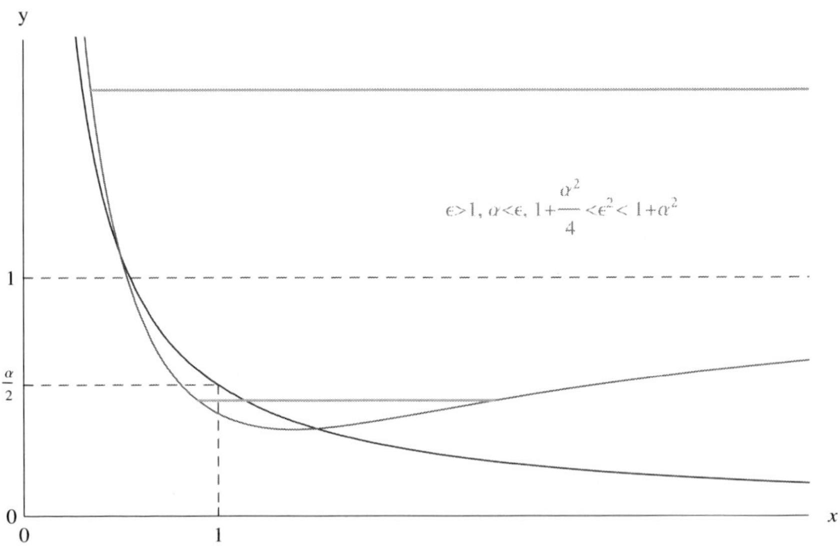

Figure 70. The effective potentials for $\alpha^2 < \varepsilon^2 < 1+\alpha^2$.

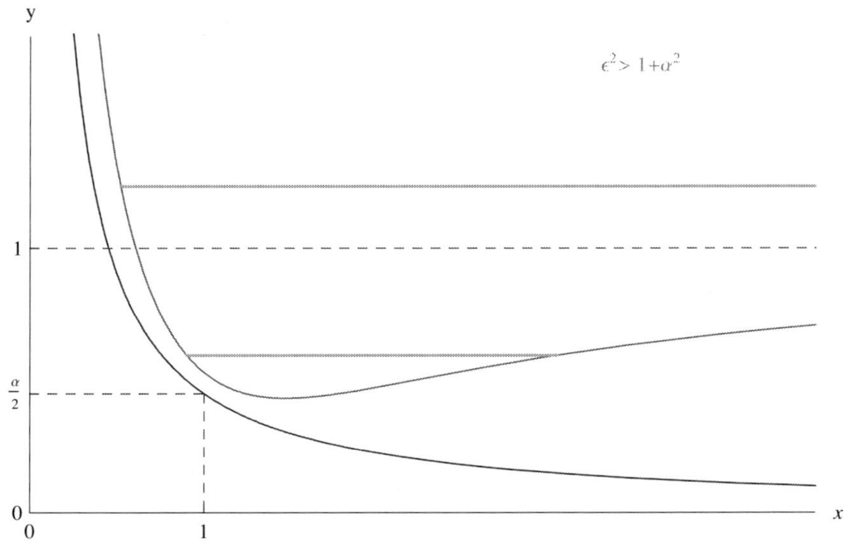

Figure 71. The effective potentials for $\varepsilon^2 > 1 + \alpha^2$.

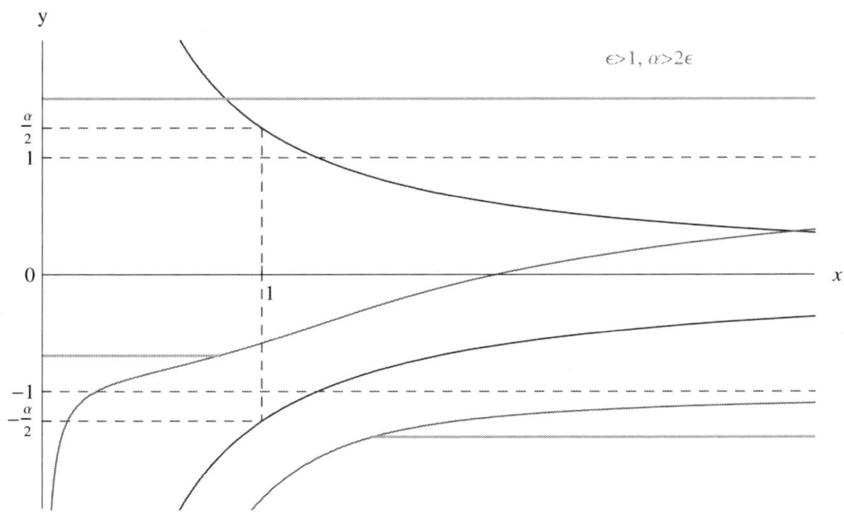

Figure 72. The effective potentials for heavy shells at $\varepsilon > 1$ and $\alpha > 2\varepsilon$.

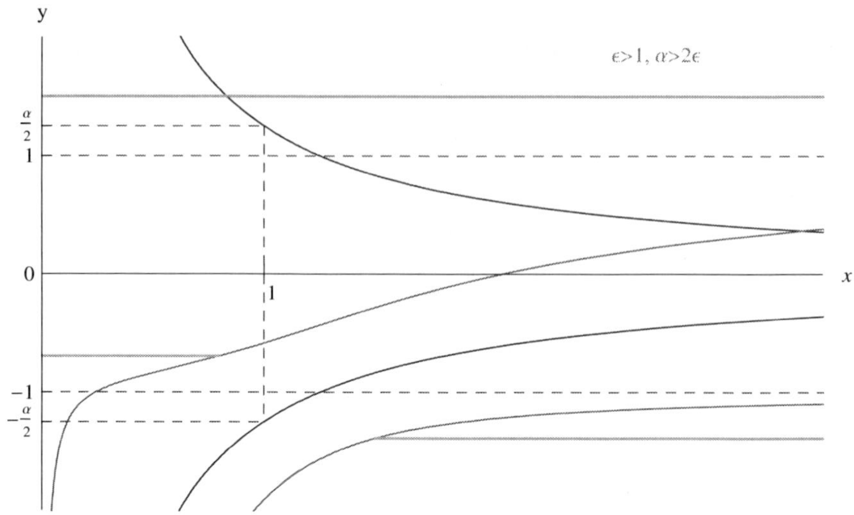

Figure 73. The effective potential for the light shells at $\varepsilon < 1$ and $\alpha < \varepsilon$.

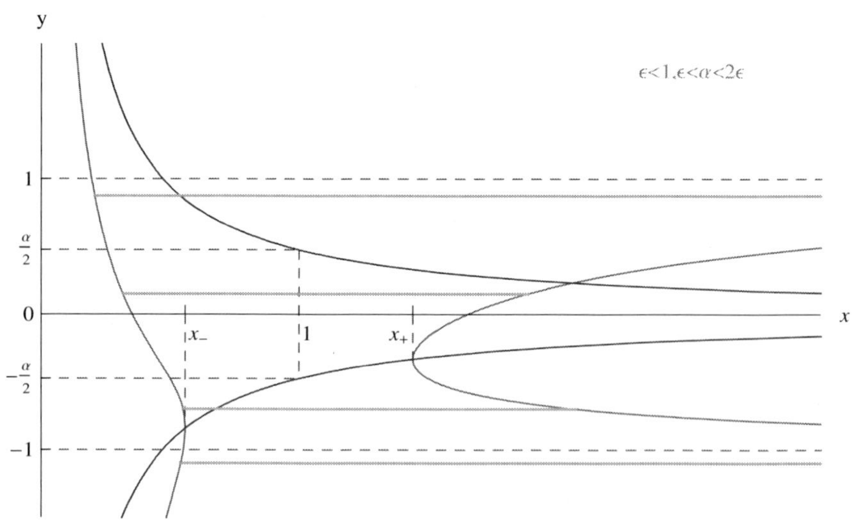

Figure 74. The effective potential for the heavy shells at $\varepsilon < 1$ and $\varepsilon < \alpha < 2\varepsilon$.

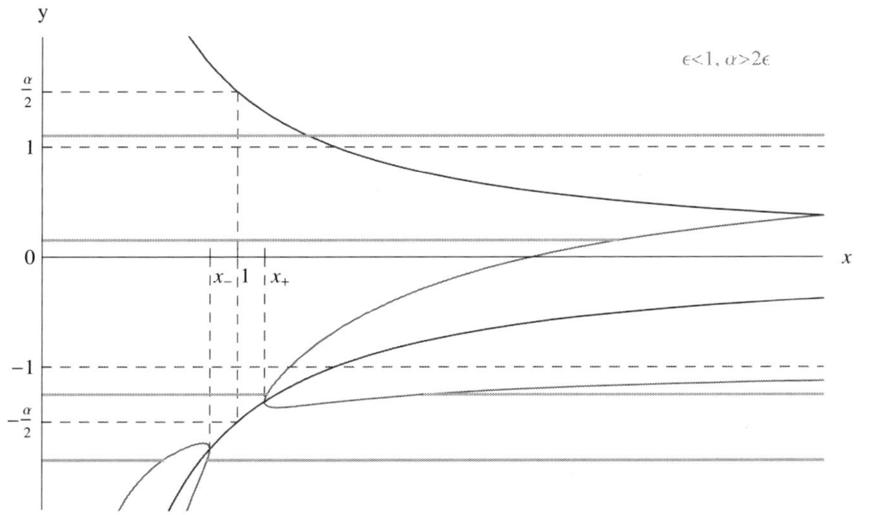

Figure 75. The effective potentials for heavy shells at $\varepsilon < 1$ and $\alpha > 2\varepsilon$.

In these notations, the apparent horizons of the inner Reissner–Nordström black holes (at $\varepsilon < 1$) are

$$x_\pm = 1 \pm \sqrt{1-\varepsilon^2}, \tag{271}$$

and we denote by x_1 and x_2 ($x_2 < x_1$) the abscissae of intersection points $y_1 = y_2$. Note, that the intersections $y_1 = \tilde{y}_2$ can exist only if $\sigma_{\rm in} = -1$, and the corresponding abscissae are simply the horizon radii, $x_\pm$.

Let us start to study the effective potential. Consider, first, its asymptotical behavior when $x \to 0$ and $x \to \infty$. If $\sigma_{\rm in} = +1$, then

$$y_1 = V_{\rm eff} \longrightarrow \frac{2\varepsilon - \alpha}{2x} - \frac{1}{\varepsilon}, \qquad x \to 0; \tag{272}$$

so, for when $x \to 0$, $V_{\rm eff} \to +\infty$, if $2\varepsilon > \alpha$, and $V_{\rm eff} \to -\infty$, if $2\varepsilon < \alpha$. For $2\varepsilon = \alpha$, $V_{\rm eff}(0) = -1/\varepsilon$. When $\sigma_{\rm in} = -1$, then always $V_{\rm eff} \to -\infty$ for $x \to 0$. At infinity, $x \to \infty$, the behavior of the effective potential is the following. If $\sigma_{\rm in} = +1$, then $V_{\rm eff} \to 1 - 0$. But, if $\sigma_{\rm in} = -1$,

$$y_1 = V_{\rm eff} \longrightarrow -1 + \frac{2-\alpha}{2x}, \qquad x \to \infty; \tag{273}$$

so, $V_{\rm eff}$ approaches $y = -1$ from above, when $\alpha < 2$, and from below, when $\alpha > 2$.

Now, how about extrema? It is easy to see that the extremum condition is

$$\sigma_{\text{in}}\sqrt{1-\frac{2}{x}+\frac{\varepsilon^2}{x^2}}=2(\frac{\varepsilon^2}{x}-1). \tag{274}$$

So, the relevant solution to this equation should obey the inequality $\sigma_{\text{in}}(\varepsilon^2/x-1)>0$. We get

$$\frac{1}{x_{\text{extr}}}=\frac{1}{\varepsilon^2}\left(1+\alpha\sigma_{\text{in}}\sqrt{\frac{\varepsilon^2-1}{4\varepsilon^2-\alpha^2}}\right). \tag{275}$$

It is clear that the extreme exist only if either simultaneously $\varepsilon^2>1$ and $4\varepsilon^2>\alpha^2$, or $\varepsilon^2<1$ and $4\varepsilon^2<\alpha$. We do not discuss here the nature of extrema, this will become quite evident while considering particular cases. Here we would like only to note, that for Reissner–Nordström black holes, when $\varepsilon<1$, $x_{\text{extr}}<x_-$ for $\sigma_{\text{in}}=+1$, and $x_{\text{extr}}>x_+$ for $\sigma_{\text{in}}=-1$.

Finally, before coming to drawing Figures, we need to consider intersections of the effective potential curve y_1 with the curves y_2 and $\tilde{y}_2(=-y_2)$. The latter is quite trivial: the intersections occur just at the horizons $x_\pm$ when they exist. Clearly, the intersections $y_1=y_2$ exist only for $\sigma=+1$. Their abscissae are

$$x_{1,2}=1\pm\sqrt{1+\alpha^2-\varepsilon^2}. \tag{276}$$

If $\varepsilon^2<\alpha^2$, we have only one point with

$$x_1=1+\sqrt{1+\alpha^2-\varepsilon^2}. \tag{277}$$

For $\alpha^2<\varepsilon^2<1+\alpha^2$ we have two intersections while for $\varepsilon^2>1+\alpha^2$ — no intersections at all. Note, that the latter may happen only for the inner Reissner–Nordström metrics with naked singularity.

We now start drawing Figures for the effective potential and begin with the case when the inner metrics is a part of the extreme Reissner–Nordström black hole, i.e., when $\varepsilon=1$. Then,

$$y_1=V_{\text{eff}}=\sigma_{\text{in}}\left|1-\frac{1}{x}\right|-\frac{\alpha}{2x}. \tag{278}$$

Though for such a metrics σ_{in} can not change its sign (it is either $+1$ everywhere, or -1 everywhere) we prefer to put both curves on the same Figure. Due to the modulus sign, each curve has a jump in the first derivative at $x=1$ (at doubled apparent horizon). Evidently, instead of drawing curves separately for $\sigma_{in} = +1$ and $\sigma_{in} = -1$, we can draw two intersecting smooth curves

$$V_1 = 1 - \frac{1}{x} - \frac{\alpha}{2x} \quad \text{and} \quad V_2 = -1 + \frac{1}{x} - \frac{\alpha}{2x}, \tag{279}$$

the resulting Figure being the same. The curves behave differently when $M/m_{in} = \alpha < 2$, such shells we will call "light" shells, and when $M/m_{in} = \alpha > 2$ — these are "heavy" shells. The effective potentials are shown in Fig. 67 and Fig. 68, respectively.

Each of the resulting Figures consists of four branches that begin at the vortex with the coordinates $(1, -\alpha/2)$. The upper branches bound the region where $\sigma_{in} = +1$, while the lower ones — the region with $\sigma_{in} = -1$. One may observe the sharp difference between the potentials for light and heavy shells. In the case of light shells the left branch for $\sigma_{in} = +1$ goes to $+\infty$ when $x \to 0$, while in the case of heavy shells it goes down to $-\infty$. Also, the potentials for light shells form the wedges, looking down when $\sigma_{in} = +1$, and up — when $\sigma_{in} = -1$, while for the heavy shells there are no wedges. The types of allowed motions are also different. The light shells have either bound trajectories with two turning points, or unbound ones with one turning point, and they never reach the naked singularity at $r = 0$. The heavy shells, on the contrary, start from the naked singularity, when $\sigma_{in} = +1$, and may have either bound trajectories with one turning point, or the unbound ones with no turning point at all. If $\sigma_{in} = -1$, the motion is always unbound with one turning point. Finally, one should distinguish between two cases: $0 < \alpha < 1$ and $1 < \alpha < 2$. In the first there are two intersection points of the effective potential $y_1 = V_{eff}$ with $\sigma_{in} = +1$ and the curve y_2 where σ_{out} changes its sign, while in the second we have the single intersection point with the right hand branch of the effective potential. In our Fig.6b we show only the latter case.

What will happen when we start to increase or decrease the electric charge $|e| = \varepsilon\sqrt{G}m_{in}$? The vortex disappears. For $\varepsilon > 1$, the curve with $\sigma_{in} = +1$ will go up, and that with $\sigma_{in} = -1$ will go down. The wedges become the minimum and maximum of the potentials for $\sigma_{in} = +1$ and $\sigma_{in} = -1$, respectively. For $\varepsilon < 1$ there forms the combined potential with two branches, left and right ones, each of them contains both the part with $\sigma_{in} = +1$ and that with $\sigma_{in} = -1$.

Let us begin with the case $\varepsilon > 1$, i.e., when the space-time inside the shell is some part of Reissner–Nordström manifold with naked singularity. First of all, we need to generalize the notions of light and heavy shells. Now, the light shells are those with $\alpha < 2\varepsilon$. For them, the effective potential is shown in Fig. 69.

In the Figure above we show the case when there is only one intersection point of the effective potential y_1 for $\sigma_{in} = +1$ with the curve y_2 where σ_{out} changes its sign. This corresponds to the double inequality $\varepsilon < \alpha < 2\varepsilon$ (surely, this requires $\alpha > 1$). In our Figure the value of the effective potential at the minimum is negative, it means that also should be $\varepsilon^2 < 1+\alpha^2/4$. Keeping α constant and increasing further ε we enter the region $\alpha^2 < \varepsilon^2 < 1+\alpha^2$ where curves y_1 and y_2 have two intersections. Note, that now it become possible to have $\alpha < 1$, $\varepsilon^2 < 1+\alpha^2/4$ as well. The effective potentials for $\alpha^2 < \varepsilon^2 < 1+\alpha^2$ and $\varepsilon^2 > 1+\alpha^2$ are shown in Figs. 70 and 71, respectively.

We did not show in these Figures the lower parts of the effective potentials corresponding to $\sigma_{in} = -1$, because they remain qualitatively the same. Note only that the lower branch (for $\sigma_{in} = -1$) has a maximum when $\alpha < 2$, and approaches at infinity the line $y = -1$ from above, and it is monotonically increases when $\alpha > 2$. The curves $\tilde{y}_2 = -y_2$ in either of the cases do not intersect the curves $y_1 = V_{eff}$, since for $\varepsilon > 1$ there are no horizons. Also, we do not discuss the special role played in our method by the points $y_1 = y_2$. This is postponed to the subsequent Sections where we will be dealing with the global geometries and Carter–Penrose diagrams.

And what about the heavy shells, $\alpha > 2\varepsilon$? Everything becomes much more simple: no minima, no maxima, just two increasing curves (for $\sigma = \pm 1$) and the curve $\tilde{y}_2 = -y_2$ in-between as is shown in Fig. 72.

Let us turn to investigation of the most interesting case, $\varepsilon < 1$, when the inner part of the complete manifold represents the Reissner–Nordström black hole. While before (for $\varepsilon \geq 1$) the value of σ_{in} was given once and forever (i.e., there may be everywhere either R_+–region with $\sigma_{in} = +1$, or R_-–region with $\sigma_{in} = -1$), now it may change its sign dynamically during the shell's evolution. The global geometry of this inner space-time is much more rich, it contains $R_\pm$- and $T_\pm$–regions and also two types of apparent horizons. As we already know, when moving from $\varepsilon = 1$ to $\varepsilon < 1$, the vortex with four branches at the doubled horizon is divided into two horizons, $x_\pm = 1 \pm \sqrt{1-\varepsilon^2}$, and for the effective potential one obtains two separate curves, left hand $(0 < x \leq x_-)$ and right hand $(x \geq x_+)$ ones, each of them contains two branches, the upper one with $\sigma_{in} = +1$ and the lower one with $\sigma_{in} = -1$. These two branches merge exactly at $x = x_-$ (on the left hand curve) and at $x = x_+$ (on the right hand curve, where the curve $\tilde{y}_2 = -y_2$ intersects the potential curves. Note that in the interval $x_- < x < x_+$ we have no potential curves at all because, in the Reissner–Nordström black holes between two horizons there lie $T_\pm$–regions where the existence of turning points with $\dot{\rho} = 0$ is impossible.

Again, one should distinguish between the light and heavy shells, with $\alpha < 2\varepsilon$ and $\alpha > 2\varepsilon$, respectively. In contrast to the case considered previously, now for the light shells the effective potential has no extrema, while for the heavy shells there is always maximum on the left hand curve and minimum on the right hand curve. The effective potentials for the light and heavy shells are shown in Fig. 73 and 74, correspondingly.

We show in this Figure the case, when there exists a minimum in the left hand branch, but it is true only when $\alpha > 2$. For $\alpha < 2$ (and still $\alpha > 2\varepsilon$) the minimum disappears, and at infinity the curve will approach the line $y = -1$ from above.

Chapter 10

Global Geometries for the Combined Systems

In the previous Section we introduced the effective potential in order to illustrate dynamics of thin shells immersed into the Reissner–Nordström manifold (which describes metrics outside an electrically charged point-like mass) and have drawn pictures for different types of its inner part (with naked singularity, extreme case and Reissner–Nordström black hole) and different values of shell's bare mass. In addition, two curves indicating the change of sign in $\sigma = \pm 1$ for inner and outer parts of the complete manifold were shown in the same pictures. It is these sign functions that determine, actually, the global geometry of the space-time under consideration. Let us remind that the allowed shell's trajectories are horizontal (green) lines $y = \Delta m/M = \mu = const$ lying above the effective potential curve $y = y_1$ for $\sigma_{\text{in}} = +1$ and below it for $\sigma_{\text{in}} = -1$. Moreover, if this horizontal (green) line goes above the corresponding curve of changing $\sigma' s$ ($y = y_2$ or $y = \tilde{y}_2 = -y_1$), then $\sigma = +1$, and $\sigma = -1$ if it goes lower. Thus, the interplay between all of them allows us to construct the Carter–Penrose conformal diagram for any set of parameters of the combined system.

The aim of the present Section is to illustrate how our method works in all qualitatively different cases. For the sake of convenience, we list below the needed notations and definitions. The inner part of the Reissner–Nordström metrics is characterized by two parameters, the mass m_{in} and the electric charge e, while its outer part — by the mass m_{out} (= the total mass of the system) and the same charge e (remember that our shell is neutral). The shell has the bare mass M, and its total mass (which includes the gravitational mass defect) equals $\Delta m = m_{\text{out}} - m_{\text{in}}$. Also, we introduced the dimensionless parameters $\mu = \Delta m/M$, $\varepsilon = |e|/(\sqrt{G}m_{\text{in}})$, $\alpha = M/m_{\text{in}}$ and the dimensionless radius of the shell $x = \rho/(Gm_{\text{in}})$. In these notations the effective potential y_1 looks as follows

$$y_1 = V_{\text{eff}} = \sigma_{\text{in}}\sqrt{1 - \frac{2}{x} + \frac{\varepsilon^2}{x^2}} - \frac{\alpha}{2x},$$

while the curves y_2 ($\tilde{y}_2 = -y_2$) where $\sigma_{\text{out}} = \pm 1$ ($\sigma_{\text{in}} = \pm 1$) change their signs are represented by

$$y_2 = \frac{\alpha}{2x}, \quad \tilde{y}_2 = -\frac{\alpha}{2x}.$$

The apparent horizons of the inner Reissner–Nordström black holes lie at $x_\pm = 1 \pm \sqrt{1-\varepsilon^2}$.

10.1. The Case $\varepsilon = 1$

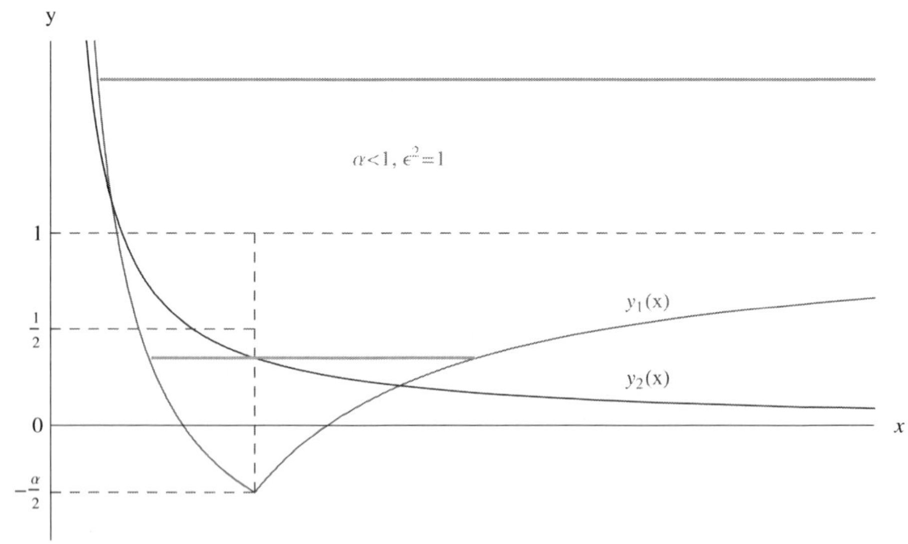

Figure 76. The effective potentials at $\varepsilon^2 = 1$ and $\alpha < 1$.

We start with the case when the inner part of complete manifold is described by the metrics of extreme Reissner–Nordström black hole, i.e., when $\varepsilon = |e|/(\sqrt{G}m_{\text{in}}) = 1$. Thus, only three parameters are left free:

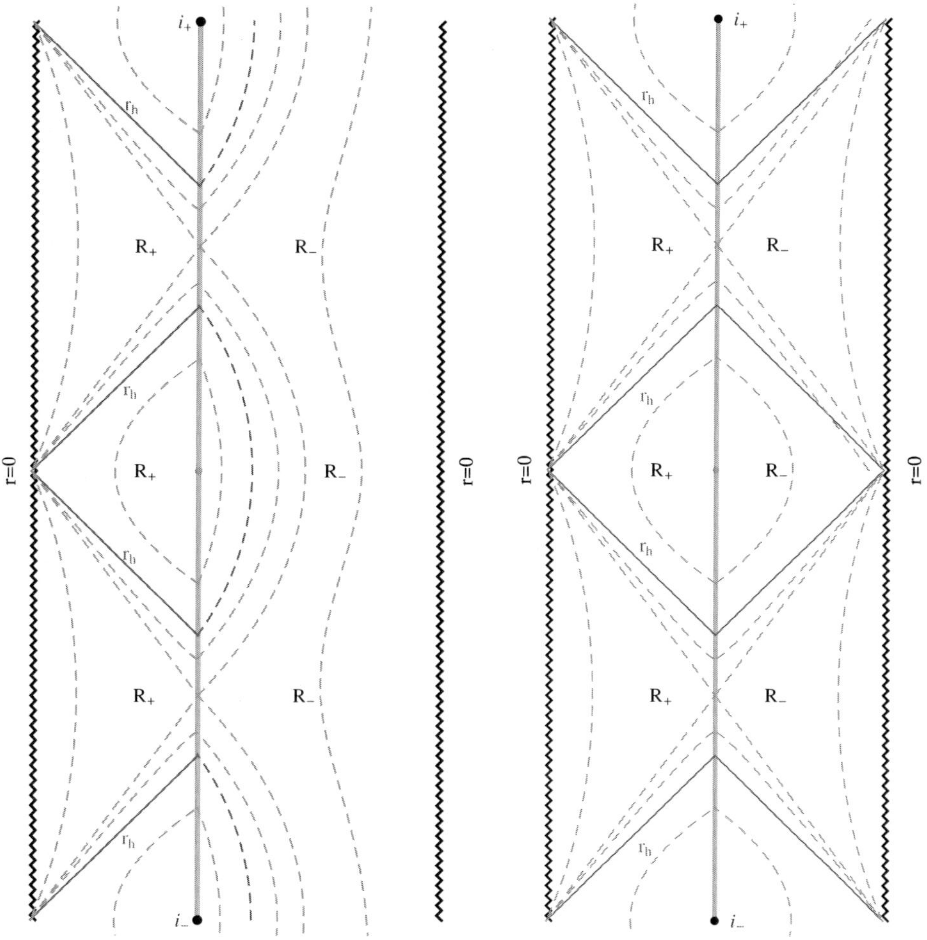

Figure 77. The combined space-time manifolds at $\varepsilon = 1$ for $\mu = \mu_1 < 0$ (left panel) and $\mu = 0$ (right panel).

1. The value of $\sigma_{\text{in}} = \pm 1$, which determines the part of the full extreme black hole space-time lying inside the shell, that one with the naked singularity ($\sigma_{\text{in}} = +1$), or with the spatial infinity ($\sigma_{\text{in}} = -1$);

2. The value of $\alpha = M/m_{\text{in}}$ which enters both the effective potential $y = y_1$ and the curves of changing $\sigma's$: $y = y_2$ for σ_{out} and $y = \tilde{y}_2 = -y_2$ for σ_{in};

3. The value of $\mu = \Delta m/M$ that determines the position of the shell's trajec-

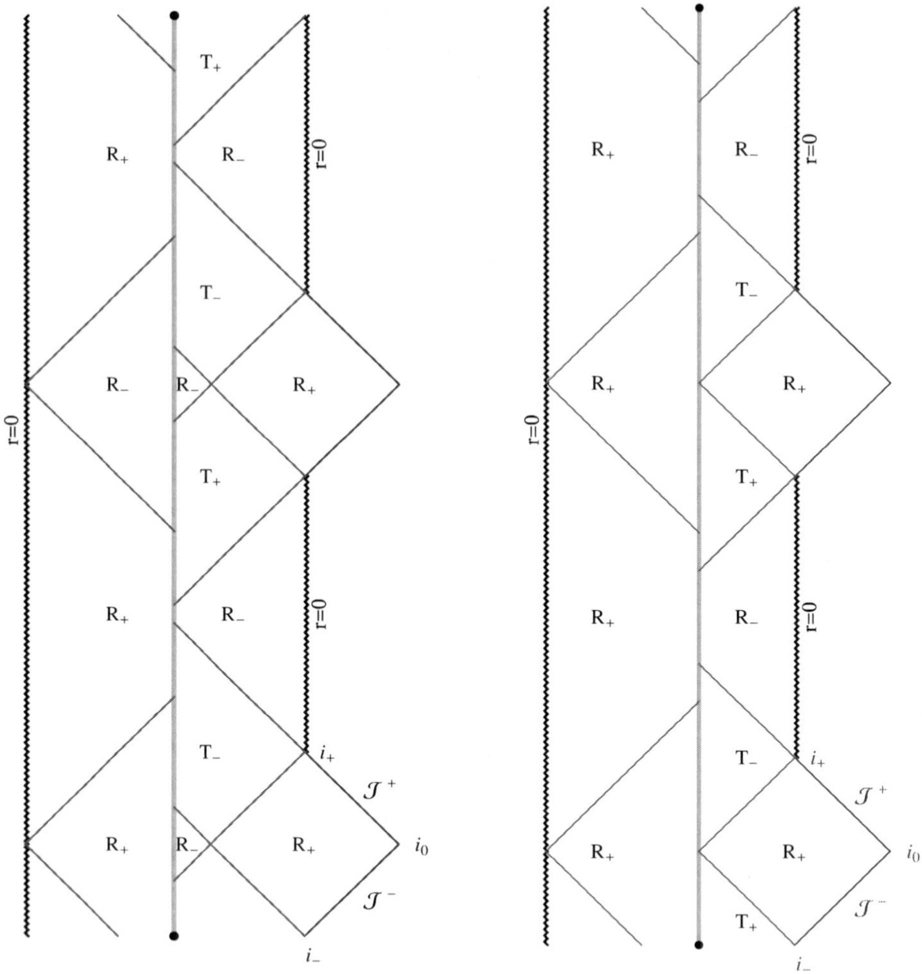

Figure 78. The combined space-time manifolds at $\varepsilon = 1$, $\alpha < 1$ at $\mu = 0$ (left panel) and $\mu = \mu_1 = \alpha/[2(1-\alpha)]$ (right panel).

tory $y = \mu$.

Consider, first, the situation with $\sigma_{\rm in} = +1$. The effective potential curve has different profiles for light shells, $\alpha < 2$, and heavy ones, $\alpha > 2$. In the case of light shells, the pictures look differently for $\alpha < 1$, when curves y_1 and y_2 have

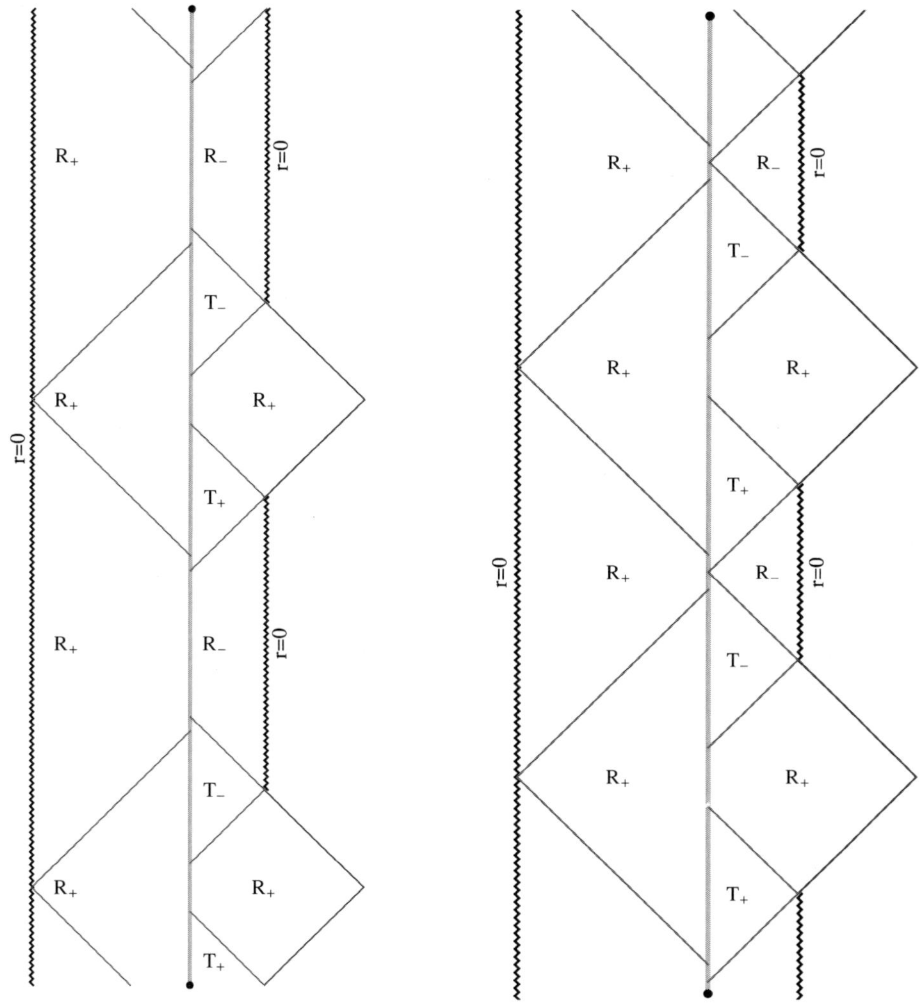

Figure 79. The combined space-time manifolds at $\varepsilon = 1$, $\alpha < 1$ and $2(1+\alpha) < \mu_3 < \alpha/[2(1-\alpha)] < 1$ (left panel) and $\mu = \mu_1 = \alpha/[2(1-\alpha)] < 1$ (right panel).

two intersection points at $x = 1 \pm \alpha$, and for $1 < \alpha < 2$, when these curves have only one common point at $x = 1+\alpha$. Moreover, if $\alpha < 2/3$, the left intersection point (at $x = 1-\alpha$) lies below the horizontal line $y = 1$ separating the bound and unbound trajectories, while if $2/3 < \alpha < 1$, then $y_1 = y_2(x = 1-\alpha) > 1$. So, our

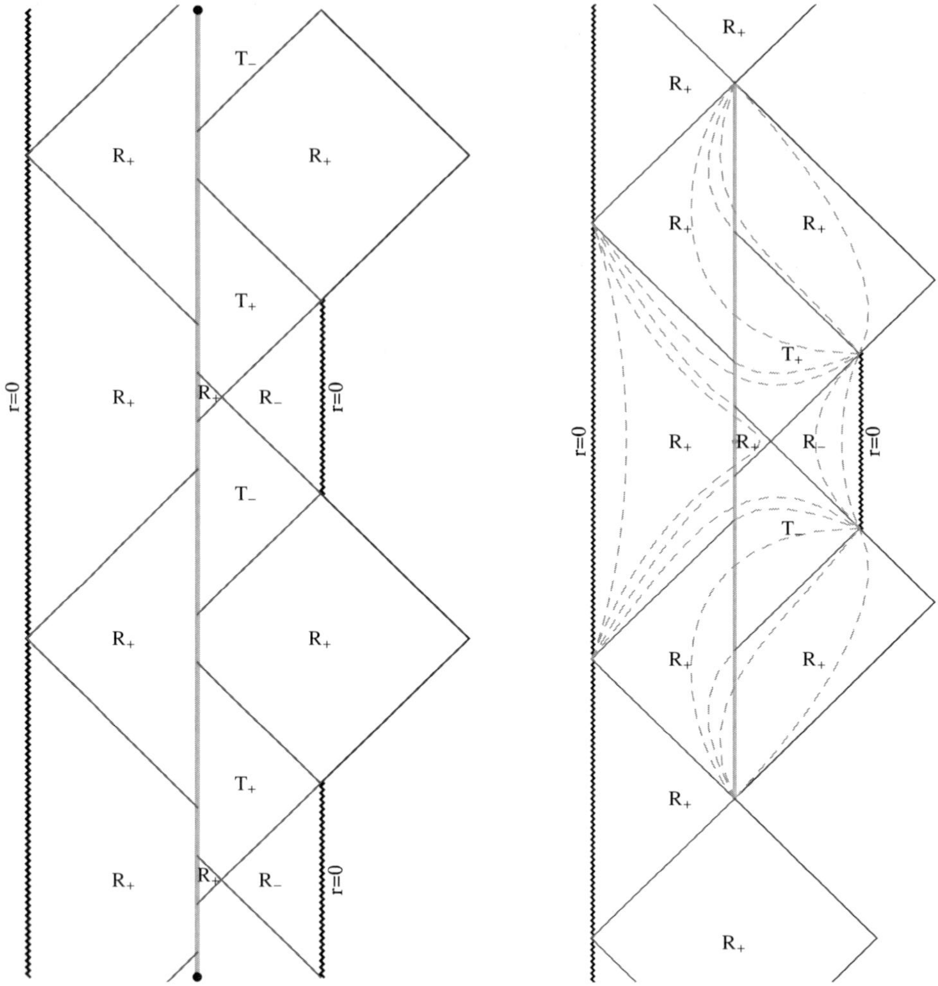

Figure 80. The combined space-time manifolds at $\varepsilon = 1$, and, respectively, at $\alpha/[2(1-\alpha)] < \mu_4 < 1$ and $\alpha < 2/3 < 1$ (left panel) and at $\mu > 1$ and $\alpha > 2/3$ (right panel).

starting point is $\varepsilon = 1$, $\sigma_{\text{in}} = +1$, $\alpha < 2/3$. The corresponding effective potential $y = y_1$ together with the curve $y = y_2$ is shown in Fig. 76. We enumerated different types of horizontal lines $y = \mu$ for further convenience. We did not draw the curve $y = \tilde{y}_2 = -y_2$ because it lies well below the present effective potential

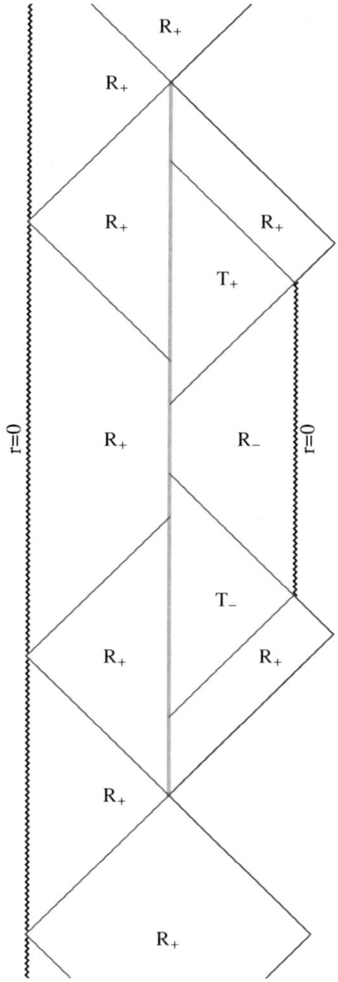

Figure 81. The combined space-time manifolds at $\varepsilon = 1$, $1 < \mu < \alpha/[2(1-\alpha)]$ and $2/3 < \alpha < 1$.

curve and, thus, does not influence the shell's trajectories. The common feature is that there are two turning points for bound and one — for unbound motion.

Let us begin to investigate the possible trajectories of the shell and construct the corresponding Carter–Penrose conformal diagrams that show us clearly the global geometry of the combined space-time manifolds. Consider, first, the

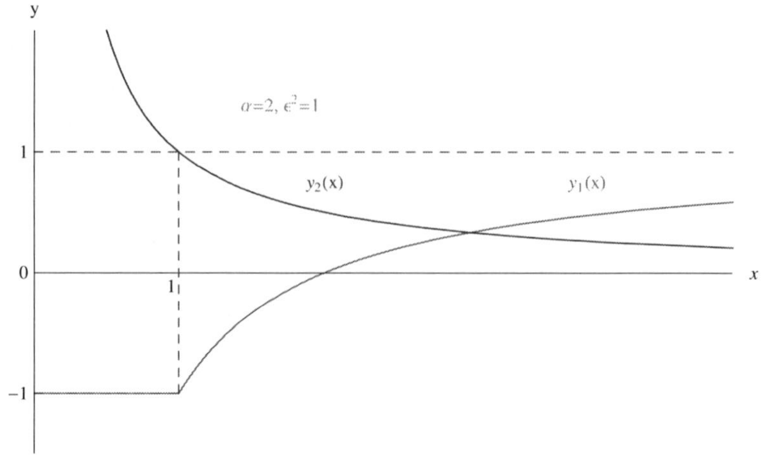

Figure 82. The effective potentials at $\varepsilon^2 = 1$, $\alpha = 2$ and $\sigma_{in} = +1$.

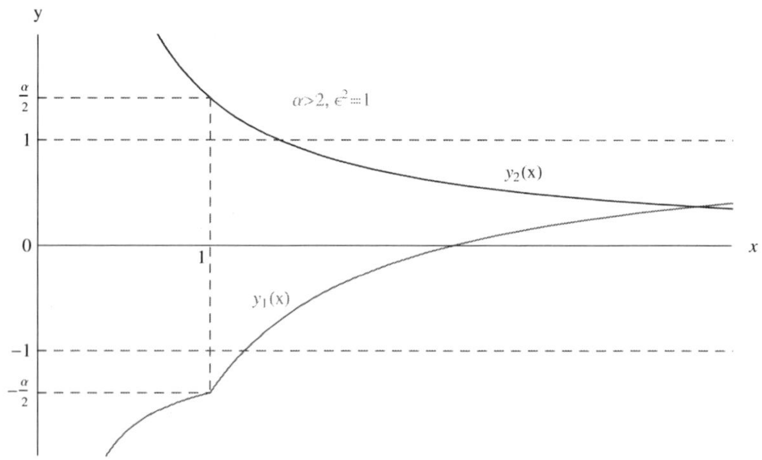

Figure 83. The effective potentials at $\varepsilon^2 = 1$, $\alpha > 2$ and $\sigma_{in} = +1$.

trajectories for which the initial conditions are such that $\mu = \mu_1 < 0$. Since $\varepsilon = 1$, the relation $m_{out} < m_{in}$ and $\Delta m < 0$ means that the outer part of the complete manifold corresponds to the case of naked singularity. Also, the horizontal line $y = \mu_1 < 0$ lies clearly below the curve $y = y_2(> 0)$, and so $\sigma_{out} = -1$. The

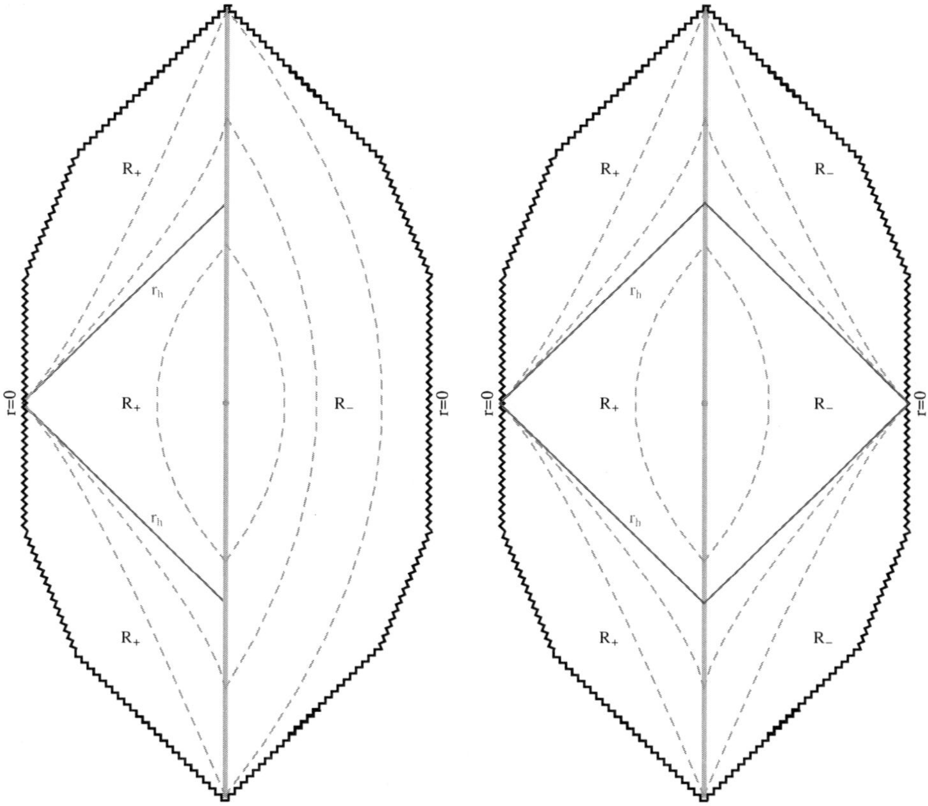

Figure 84. The combined space-time manifolds at $\alpha > 2$ and $< \mu < 0$ (left panel) and $< \mu = 0$ (right panel).

Carter–Penrose diagram for this case is shown in Fig. 77 (left panel). The shell's motion is bound with two turning points. On the diagram the left hand turning point lies between the singularity at zero radius and the doubled horizon, while the right hand one is beyond the horizon. By dashed curves we indicated the surfaces of constant radii, this helps better understanding of the global geometry.

In the limiting case $\mu = 0$ the Carter–Penrose diagram is left-right symmetric (see the right panel in Fig. 77). Both inside and outside the shell we have the extreme Reissner–Nordström black hole.

Now, let us turn to the positive values of μ and consider the case when

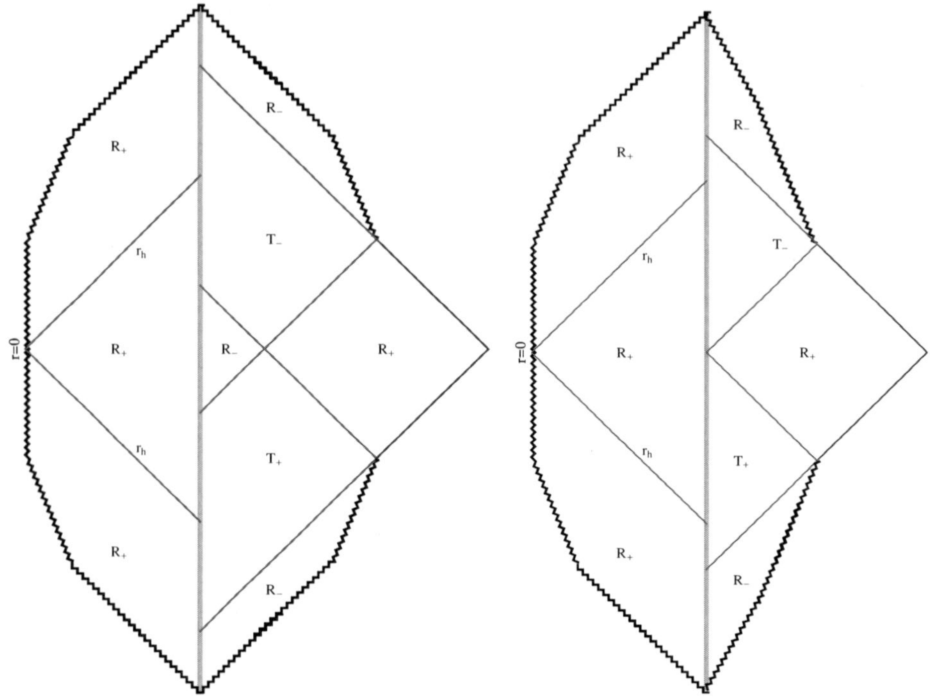

Figure 85. The combined space-time manifolds at $\alpha > 2$ and $\mu < \alpha/[2(1+\alpha)]$ (left panel) and at $\alpha > 2$ and $\mu = \alpha/[2(1+\alpha)]$ (right panel).

the horizontal line $y = \mu$ lies below the right hand (lower) intersection point $y_1 = y_2$, i. e., $0 < \mu_2 < \alpha/[2(1+\alpha)]$. In this case $\sigma_{\mathrm{out}} = -1$ everywhere on the trajectory. Since with μ increasing from zero value, the turning points x_1 and x_2 goes, respectively, to the left and to the right off the primarily doubled horizon, the continuity properties of the relevant expressions require that $x_1 < x_-$ and $x_2 < x_+$, until we reach the level $\mu = y_1 = y_2$ at $x = x_2$ when this turning point lies exactly at the event horizon x_+ of the outer metrics. So, for $\mu = \mu_2$, the right hand turning point lies in the R_-–region of the outer metrics outside the event horizon $(x = x_+)$ (with asymptotically flat infinity), while the left hand one — in the R_-–region beyond the inner (Cauchy) horizon $x = x_-$ near the singularity at zero radius. The corresponding conformal diagram with the shell's trajectory (in light-blue) for the combined manifold is shown in Fig. 78 (left panel).

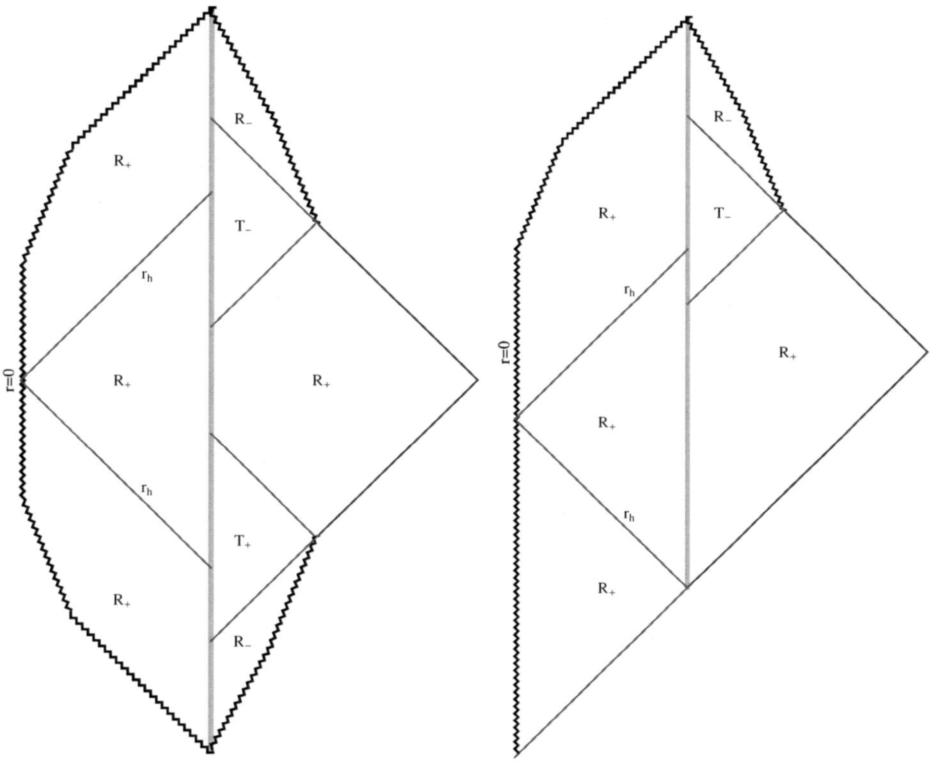

Figure 86. The combined space-time manifolds at $\alpha > 2$ and $\alpha/[2(1+\alpha)] < \mu < 1$ (left panel) and at $\alpha > 2$ and $\mu > 1$ (right panel).

When $\mu = \alpha/[2(1+\alpha)]$, the right hand turning point $(x = x_2)$ coincides with the event horizon $x = x_+$ (just at the bifurcation point) of the Reissner–Nordström metrics outside the shell. In the case of $2(1+\alpha) < \mu_3 < \alpha/[2(1-\alpha)] < 1$, where the last inequality holds because of $\alpha < 2/3$, the shell starts, say, from one of the turning points x_2 in the R_+–region $(\sigma_{\text{out}} = +1)$ of the outer metrics outside the event horizon x_+. This shell collapses and goes through the T_-–region between two horizons, where σ_{out} changes sign, then crosses the inner horizon x_- and enters the R_-–region $(\sigma_{\text{out}} = -1)$ near the singularity and reaches afterwards the another turning point x_1. When $\mu = \alpha/[2(1-\alpha)]$, this second turning point lies exactly at the inner horizon $x = x_1 = x_-$. The corresponding Carter–Penrose diagrams are shown in Figs. 78 and 79.

The trajectories with $\alpha/[2(1-\alpha)] < \mu_4 < 1$ differ from that of μ_3-type in that the turning point $x_1\,(< x_-)$ lies now in the R_+–region near the singularity of the outer metrics¡ and $\sigma_{out} = +1$ everywhere. The global geometry is shown in Fig. 80 (left panel).

If $\mu > 1$, the motion of the shell is unbound, we denoted it as of μ_5-type. The shell starts to collapse from the past temporal infinity in the R_+–region of the outer metrics and comes through the T_-–region to the R_+–region near the singularity at zero radius where there lies the single turning point $x_1\,(< x_-)$, and it then expands through the T_+–region into the R_+–region outside the event horizon (x_+) (which is different from the primary one) and, finally, the motion ends at the future temporal infinity. The Carter–Penrose diagram is finite in time for the outer part, but still has the infinite ladder structure for its inner part. It is shown in Fig. 80 (the right panel) with the corresponding level-lines $r = const$.

Let now be $2/3 < \alpha < 1$. What is changing? The left hand intersection point moves up, $y_1 = y_2 > 1$, so, the μ_4-type trajectories for bound motion disappear. Instead, the unbound motion with the left hand turning point $x_1\,(< x_-)$ in R_-–region near the singularity becomes possible for $1 < \mu < \alpha/[2(1-\alpha)]$. The Carter–Penrose diagram for this case is shown in Fig. 81. All other types of the diagrams were already given, and it is needless to repeat drawing.

If $1 < \alpha < 2$, the situation is even more simple. The curves y_1 and y_2 have only one intersection — with the right hand branch of the effective potential. So, for $\mu > \alpha/[2(1+\alpha)]$ all the trajectories will undergo the change of σ_{out}, and the left hand turning point $x_1\,(< x_-)$ lies in the R_-–region of the outer metrics near singularity. Again, all the corresponding diagrams were drawn before.

We are coming now to investigation of heavy shells, when $\alpha > 2$. There is no more a decreasing branch of the effective potential, so, there can be only one turning point for bound motion and no turning points at all for unbound motion. The shells are moving from the singularity at zero radius, then either reach the turning point and collapse back, or escape to infinity. In the limiting case $\alpha = 2$ the picture for the curves y_1 and y_2 is shown in Fig. 82, while the case corresponding to $\alpha > 2$ is shown in Fig. 83.

We confine ourselves to positive values of the system's total mass m_{out}. In our notations this means that $\mu > -1/\alpha > -1/2$. Thus, as it can be easily seen in the picture above, there are four different types of allowed shell's trajectories: $-1/\alpha < \mu_1 < 0$, $0 < \mu_2 < \alpha/[2(1+\alpha)]$, $2(1+\alpha) < \mu_3 < 1$ and $\mu_4 > 1$. We are not going to discuss the details here, it was already done before. We just show the corresponding Carter–Penrose diagrams one by one in Figs. 84-86.

Meanwhile, two notes are in order: (1) at the singularity at zero radius (which is the starting point of all the heavy shell trajectories) we always have $\sigma_{out} = -1$; (2) in the case of unbound motion we show only the collapsing shell, while the time reversal picture is also possible.

We investigated all possible global geometries in the situation when the inner part of the complete manifold represents the extreme Reissner–Nordström black hole with $\sigma_{in} = +1$. And how about $\sigma_{in} = -1$? In such a case there is no center (with zero value of radius) inside the shell, but instead, there are infinities (spatial, null and temporal ones). Again, the pictures of the effective potential are different for light and heavy shells. We would like to remind that now for the allowed trajectories the corresponding parts of the horizontal lines $y = \mu$ should lie below the curves $y = y_1$. As it was done before, we begin with the light shells, $\alpha < 2$. We see that there are two types of the trajectories: for bound motion with two turning points if $-1 < \mu < -\alpha/2$, and that ones for unbound motion if $\mu < -1$. The Carter–Penrose diagrams are shown in Fig. 87.

For heavy shells, $\alpha > 2$, the effective potential looks as follows in Fig. 88. Evidently, there can exist only unbound trajectories with one turning point when $\mu < -\alpha/2 < -1$. The conformal diagram is the same as for the light shells in the case of unbound motion.

10.2. The Case $\varepsilon > 1$

In the present Section we investigate the case when the inner part of combined space-time manifold is represented by the Reissner–Nordström metrics with naked singularity, i.e., when $\varepsilon > 1$. Here, as before, we will consider separately the cases $\sigma_{in} = +1$ and $\sigma_{in} = -1$. We already know that if $\sigma_{in} = +1$, the effective potential curve $y = y_1 = V_{eff}$ behaves differently at $x \to 0$ for "light" shells, $\alpha < 2\varepsilon$, and "heavy" ones, $\alpha > 2\varepsilon$. In the case of light shells the minimal value of the effective potential may be both negative and positive, and the interplay between the curves y_1 and y_2 (which is of most importance in constructing the global geometries) becomes rather tricky.

Of course, we begin with the light shell, $\alpha < 2\varepsilon$. Let us fix the value of α and gradually increase ε, starting from $\varepsilon = 1$. In the course of doing this we meet three key points:

1. $\varepsilon_1^2 = \alpha^2$, when there appears the second (left) intersection point of the curves y_1 and y_2;

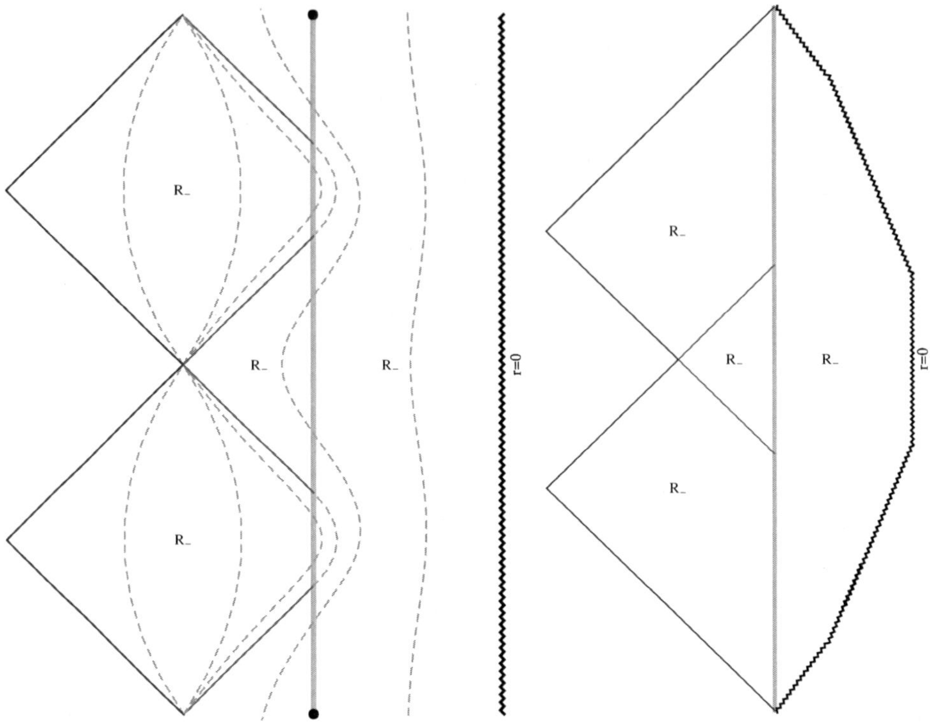

Figure 87. The combined space-time manifolds for the light shells at $\alpha < 2$ and, respectively, $-1 < \mu < -\alpha/2$ (left panel) and $\mu < -1$ (left panel).

2. $\varepsilon_2^2 = 1 + \alpha^2/4$, when the minimum of the effective potential crosses the abscissae axis;

3. $\varepsilon_3^2 = 1 + \alpha^2$, when two intersection points $y_1 = y_2$ merges. Note that the order of the first two points may be reversal ($\varepsilon_1 < \varepsilon_2$ if $\alpha^2 < 4/3$).

We start our consideration from the case $\alpha < 1 (< \varepsilon < 2\varepsilon)$. Then, for $\sigma_{in} = +1$ and ε slightly greater than 1, the picture of the curves $y_1 = V_{eff}$ and y_2 is shown in Fig. 89.

Here we deliberately chose the case when the effective potential is negative at the minimum in order to show that such a case is, in principle, may take place. Let us investigate it in details and construct all the possible global geometries. We start with the trajectories of μ_1-type. By this we denoted all the cases. when there is a naked singularity in the outer metrics. The value of μ_1 can be both

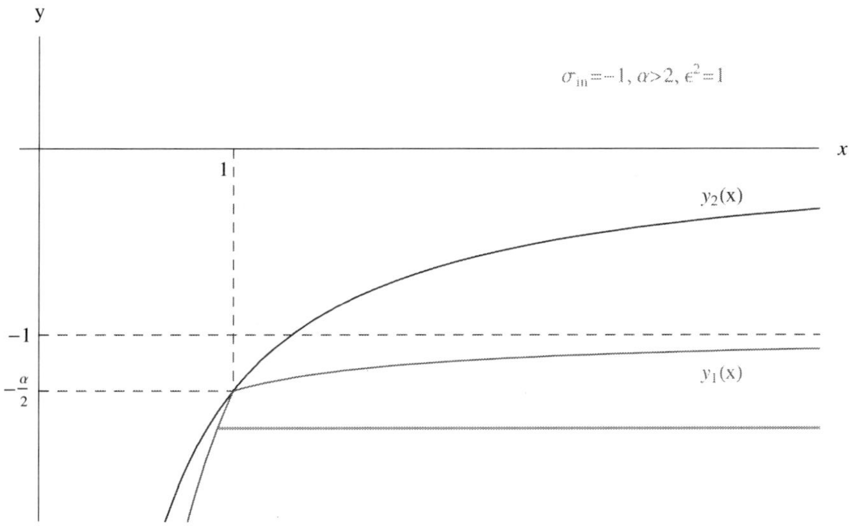

Figure 88. The effective potential for heavy shells at $\alpha > 2$, $\varepsilon^2 = 1$ and $\sigma_{in} = -1$.

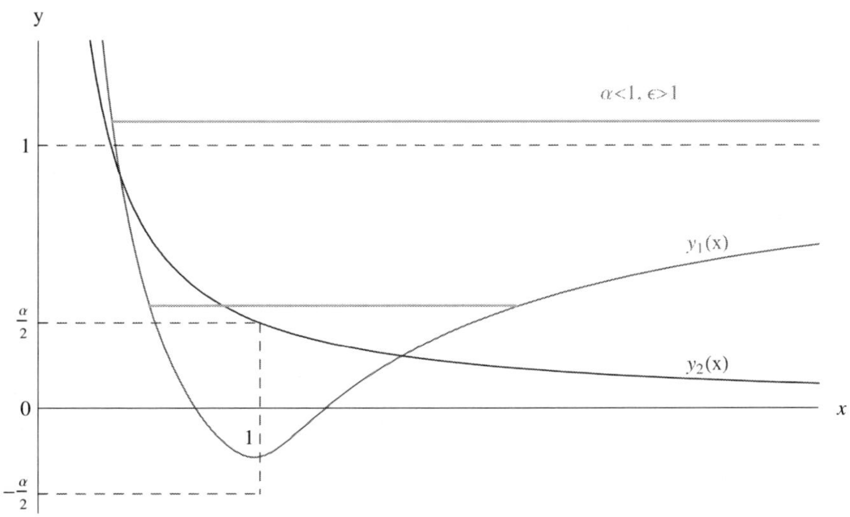

Figure 89. The effective potential at $\alpha < 1 (< \varepsilon < 2\varepsilon)$.

negative (but, of course, greater than the minimum of the potential) and positive, but less than $\mu = (\varepsilon - 1)/\alpha$, when the outer metrics becomes that of the extreme Reissner–Nordström black hole (note, that $\sigma_{out} = -1$). The Carter–Penrose diagram looks as follows in Fig. 90 (left panel).

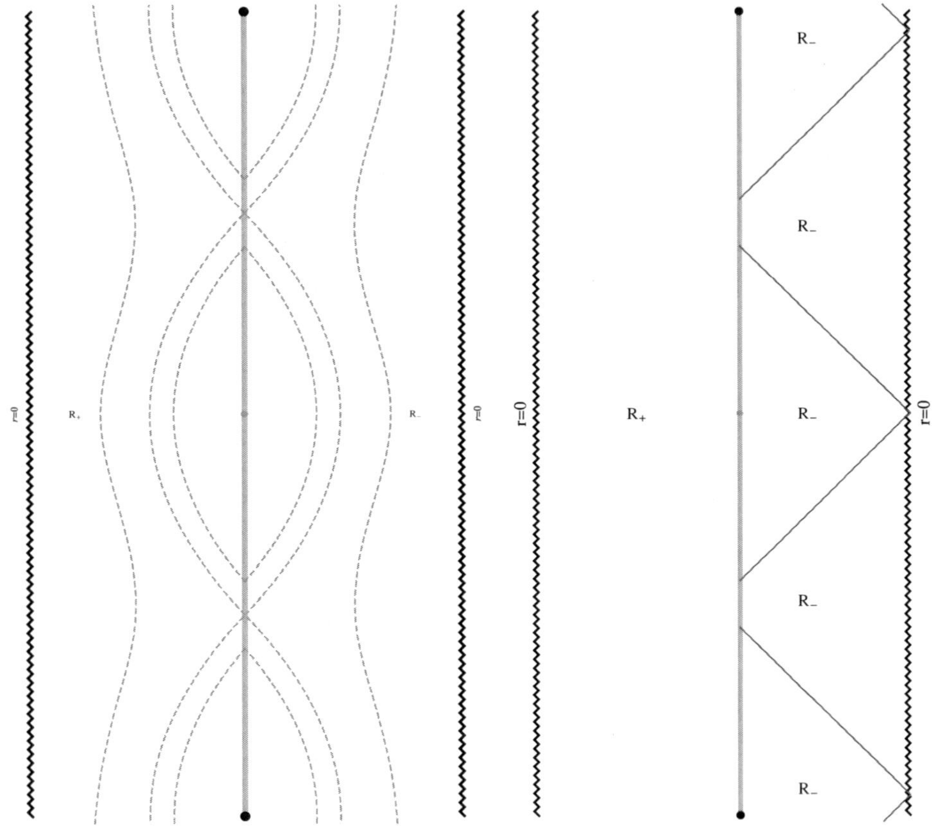

Figure 90. The Carter–Penrose diagram at $\alpha < 1(< \varepsilon < 2\varepsilon)$, $\sigma_{in} = +1$ and, $\sigma_{out} = -1$ at, respectively, $\mu < (\varepsilon - 1)/\alpha$ (left panel) and $\mu = (\varepsilon - 1)/\alpha$ (right panel).

In the limiting case $\mu = (\varepsilon - 1)/\alpha$ the conformal diagram is shown in Fig. 90 (right panel).

The next step is the μ_2-type trajectories. Since $\mu_2 > (\varepsilon - 1)/\alpha$, the outer metrics is that of the Reissner–Nordström black hole, but we have it still lower than the right hand intersection point $y_1 = y_2$, so everywhere along the trajectory $\sigma_{out} = -1$. The corresponding conformal diagrams are shown in Fig. 91.

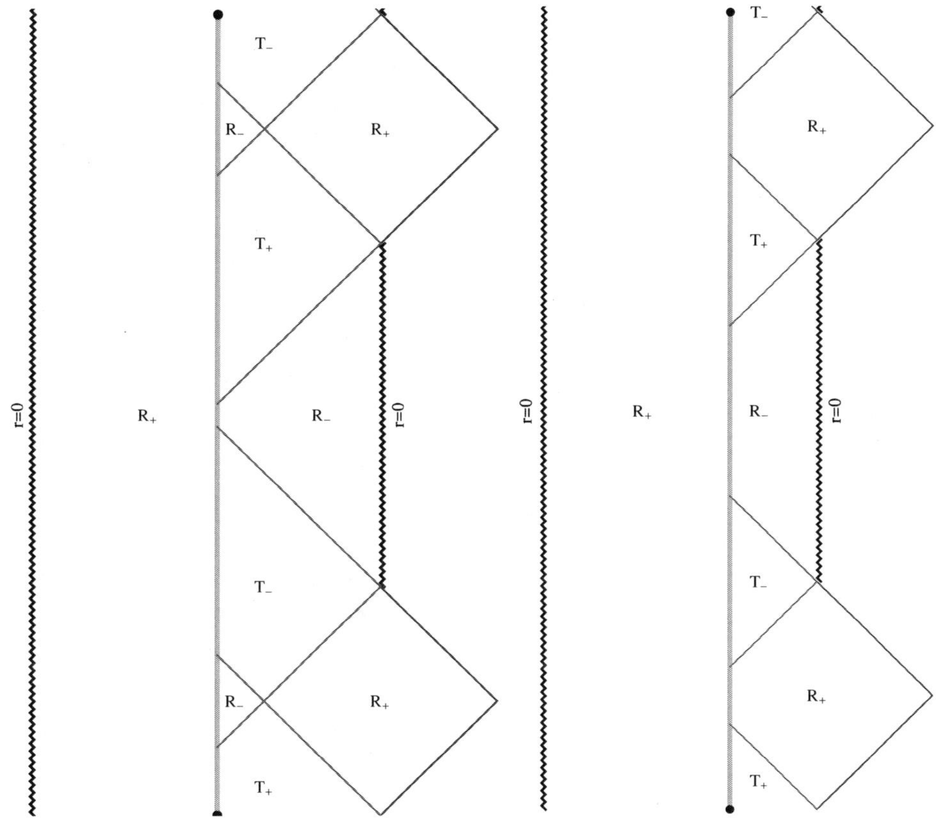

Figure 91. The Carter–Penrose diagram at $\alpha < 1 < \varepsilon$ at, respectively $(\varepsilon - 1)/\alpha < \mu < \mu_2$ (left panel) and $\mu_2 < \mu < \mu_1$ (right panel).

Increasing μ further, we come to the μ_3-type trajectories when they start from the left hand turning point with $\sigma_{out} = -1$, (R_-–region near singularity), go through the T_+–region where σ_{out} changes its sign, enter the R_+–region outside the event horizon ($\sigma_{out} = +1$) and then either meet the second (right) turning point, or escape to infinity. The relevant Carter–Penrose diagramsfor bound and unbound motions are shown in Fig. 92.

Finally (for our specific values of α and ε), the μ_4-types of the trajectories,

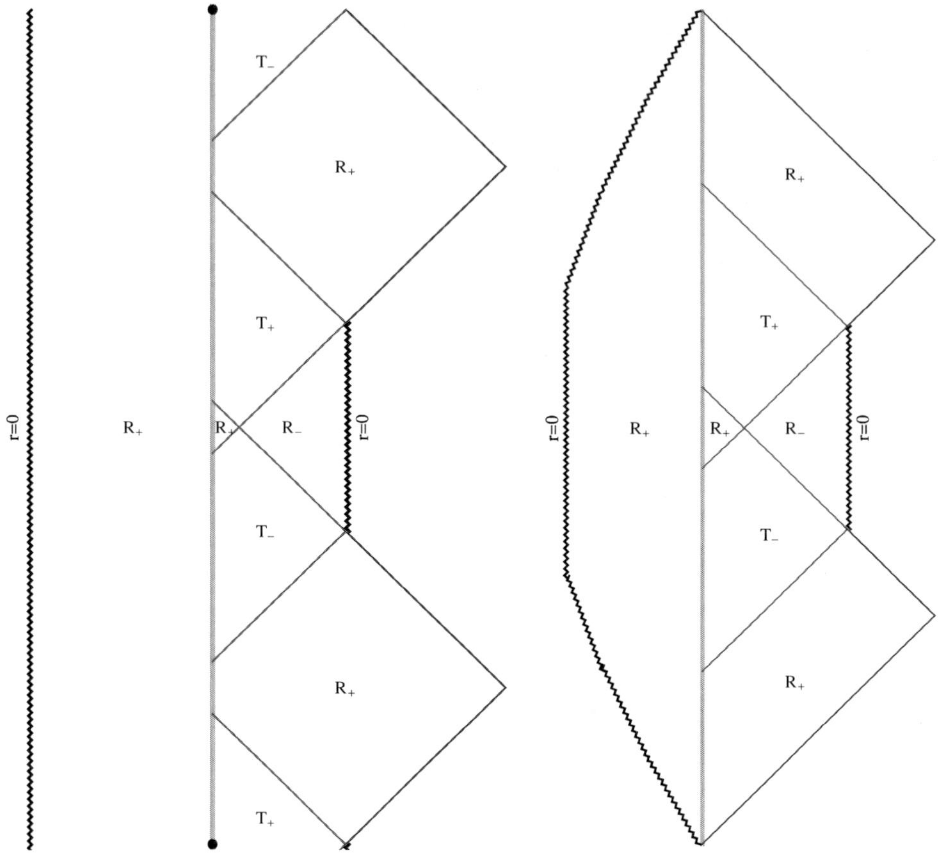

Figure 92. The Carter–Penrose diagram at $\alpha < 1 < \varepsilon$, respectively, for bound motion at $\mu_1 < \mu < 1$ (left panel) and for unbound motion at $\mu > 1$ (right panel).

when $\sigma_{\rm out} = +1$ everywhere. Again, there can be both bound and unbound motions. See below the corresponding conformal diagrams in Fig. 93.

Keeping $\alpha < 1$, let us increase ε and enter the layer $1+\alpha^2/4 < \varepsilon^2 < 1+\alpha^2$, where there are two intersection points $y_1 = y_2$. But, one should distinguish between two cases: $\alpha^2 > 2\varepsilon(\varepsilon-1)$, when the intersection points lie on different (left-right, decreasing-increasing) branches of the effective potential, and $\alpha^2 < 2\varepsilon(\varepsilon-1)$, when both of them are on the same (left-decreasing) branch. The sets of possible global geometries are quite different in these two cases. To understand the problem better one should consider the triple intersection points $y=\mu=y_1=y_2$, where the turning points of the trajectories lie exactly at the horizons of the outer metrics. The question is: which of the horizons, inner (Cauchy) or outer (event) ones? It appears that when the intersection points are on different branches of the effective potential, then the lower (right hand) one corresponds to the turning point at the event horizon and the upper (left hand) one - to the Cauchy horizon. If the intersection points are on the same (decreasing) branch, both of them correspond to the turning points at the inner horizons (surely, different ones due to different values of $\mu(m_{\rm out})$. Note, that if $\alpha^2=2\varepsilon(\varepsilon-1)$, the right hand intersection point lies exactly at the minimum of the effective potential. Moreover, the triple point in such a case corresponds to the extreme Reissner–Nordström black hole outside the shell (i.e., $m_{\rm out}/m_{\rm in}=\varepsilon$). Consider, first, the smaller value of electric charge, when $\alpha^2>2\varepsilon(\varepsilon-1)$. The effective potential curve y_1 together with the curve $y=y_2$ is shown in Fig. 93.

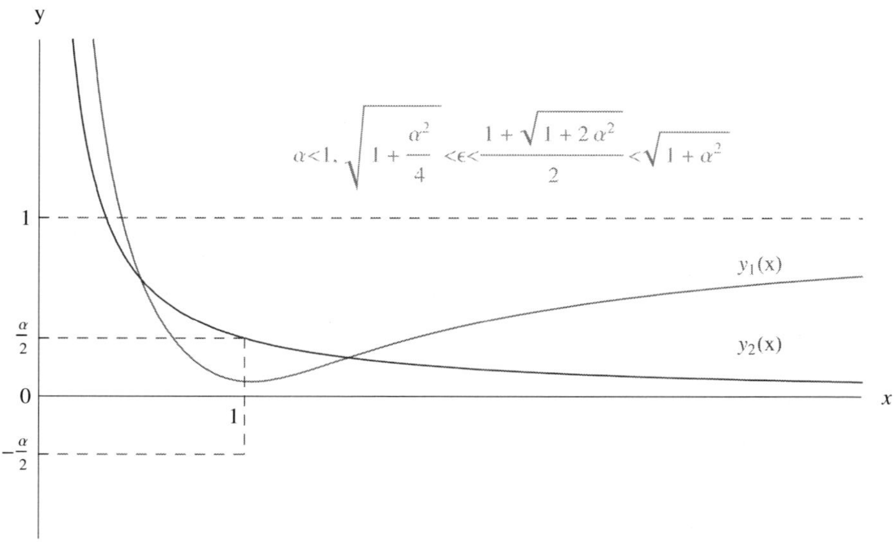

Figure 93. The effective potential for heavy shells at $\alpha<1$, $1+\alpha^2/4<\varepsilon^2<1+\alpha^2$ and $\alpha^2>2\varepsilon(\varepsilon-1)$.

Evidently, in what concerns the global geometries, there is nothing new compared to the previous results. And we will not repeat drawings. Now, let us come to the case $\alpha^2<2\varepsilon(\varepsilon-1)$. The effective potential curve $y=y_1$ together

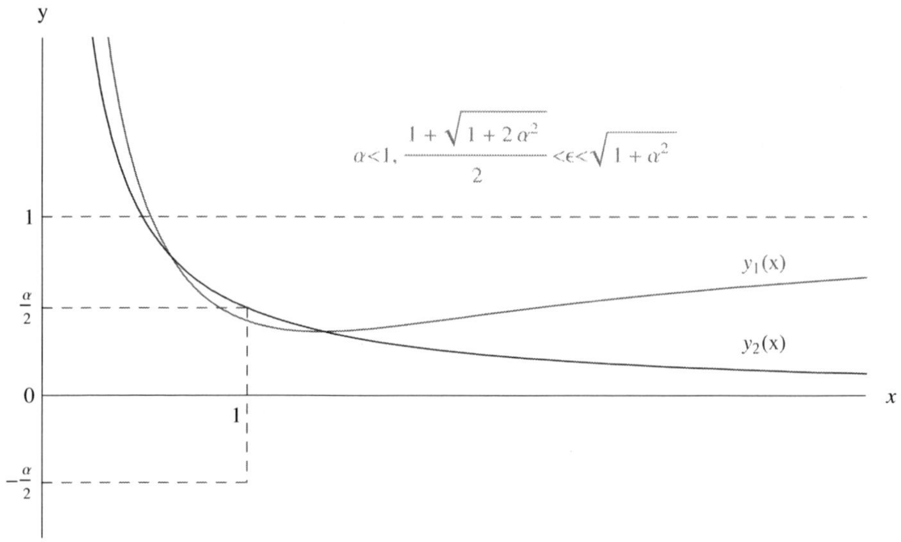

Figure 94. The effective potential for heavy shells at $\alpha < 1$, $1+\alpha^2/4 < \varepsilon^2 < 1+\alpha^2$ and $\alpha^2 < 2\varepsilon(\varepsilon-1)$.

with the curve $y = y_2$ take now the form, shown in Fig. 94.

If the horizontal line $y = \mu$ lies between the minimum of the effective potential and the lower (which is now to the left) intersection point $y_1 = y_2$, then it is possible (not, of course, always, but for certain intervals of parameters) to have the Reissner–Nordström naked singularity or the extreme black hole in the outer metrics and the shell's trajectories with $\sigma_{out} = +1$ everywhere. The Carter–Penrose diagrams in these two case are shown in Fig. 95. All other possible global geometries were already illustrated before.

Let now $\varepsilon^2 > 1+\alpha^2$. In this case the curve $y = y_2$ is lower than the potential curve $y = y_1$, so for all the trajectories $\sigma_{out} = +1$ everywhere. What is new? The main thing is that the horizontal line $y = (\varepsilon-1)/\alpha$ indicating the extreme black hole in the outer metrics can lie above the line $y = 1$ (the threshold between the bound and unbound motions), therefore, when such situation occurs, in the case of the extreme black hole or naked singularity (in some interval of ε) the only possible type of motion is the unbound one. These diagrams that are shown below in Fig. 96.

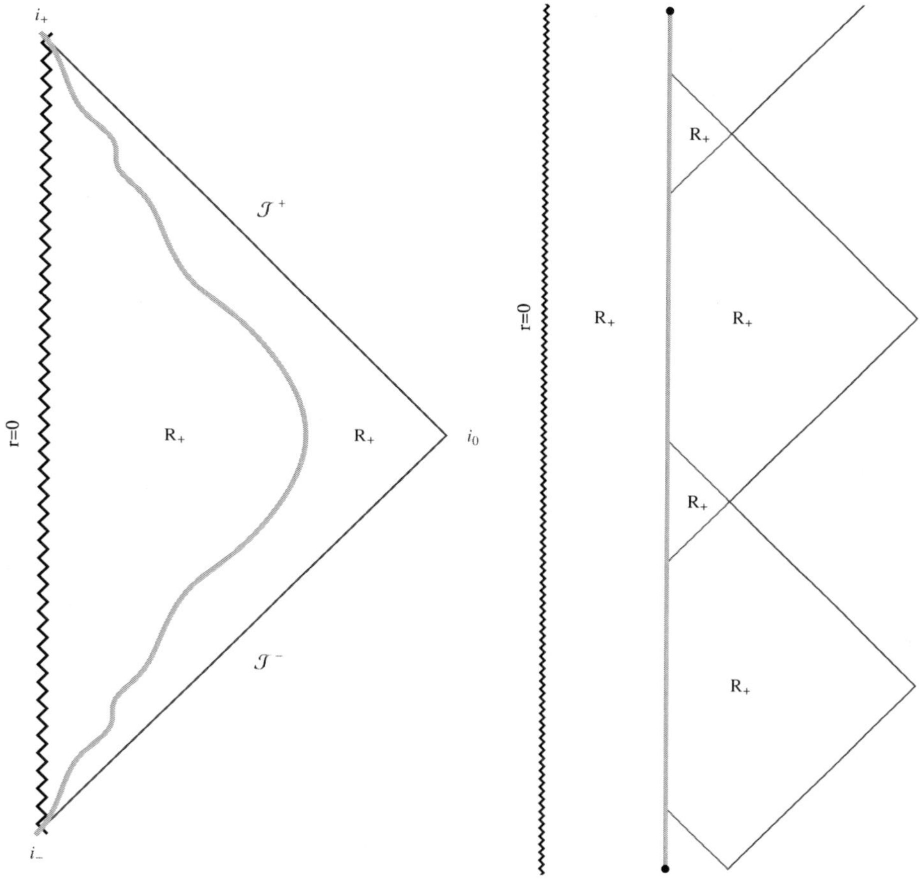

Figure 95. The Carter–Penrose diagram at $\alpha < 1$, $\sigma_{out} = +1$ and $(1 + \sqrt{1+2\alpha^2})/2 < \varepsilon < 1+\alpha^2$ at, respectively, $\mu < (\varepsilon - 1)/\alpha$ (left panel) and $\mu = (\varepsilon - 1)/\alpha$ (right panel).

What will happen when we shift the parameter α to $1 < \alpha < 2\varepsilon$? There appears the interval for ε, namely, $1 < \alpha < 2\varepsilon$, for which $y_2 > y_1$ $(x \to 0)$, and, therefore, we have only one intersection point $y_1 = y_2$. The picture for these two curves is shown in Fig. 97.

There are no qualitatively new Carter–Penrose diagrams (i. e,, the global geometries) in this case.

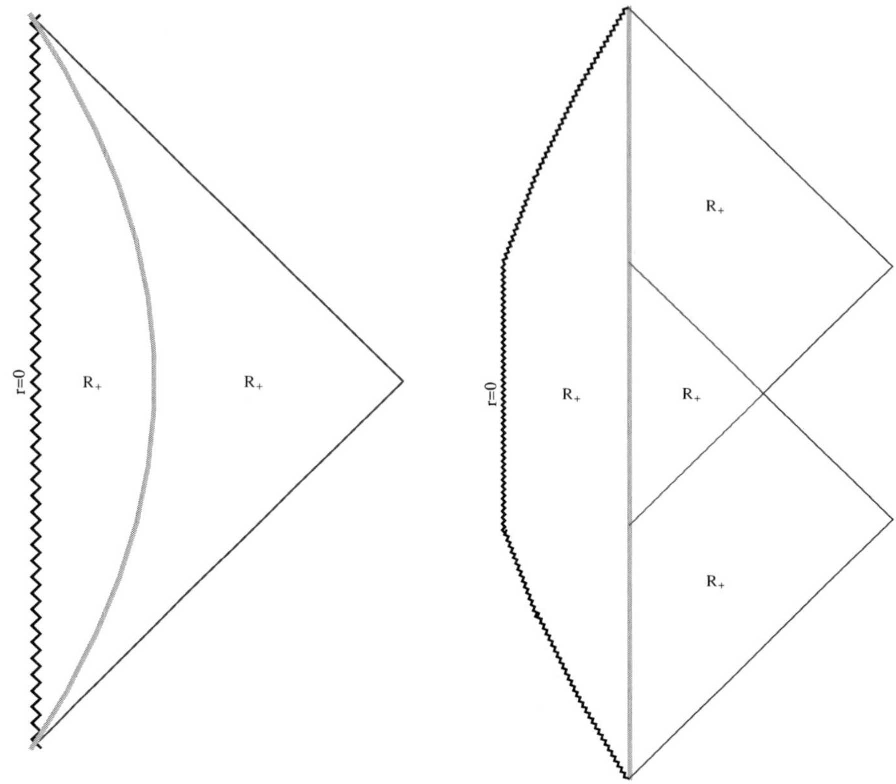

Figure 96. The Carter–Penrose diagram at $\alpha < 1, \varepsilon > 1+\alpha^2, 1 < \mu < (\varepsilon-1)/\alpha$ and (left panel) and $1 < \mu = (\varepsilon-1)/\alpha$ (right panel).

What concerns the heavy shells, $\alpha > 2\varepsilon$, the situation is qualitatively the same as in the case $\varepsilon = 1$. One should only remove the (red) lines indicating the doubled horizons in the inner part of the Carter–Penrose diagrams, Figs. 84—86.

Finally, let us consider the case $\sigma_{\rm in} = -1$, when inside the shell there is singularity at zero radius but, instead, the infinities (spatial, null and temporal). The effective potential curves look differently for $\alpha < 2\varepsilon$ (light shells) and for $\alpha > 2\varepsilon$ (heavy shells). They are shown in Figs. 98 and 99, respectively.

Remember, that now the allowed trajectories (the horizontal lines $y = \mu$)

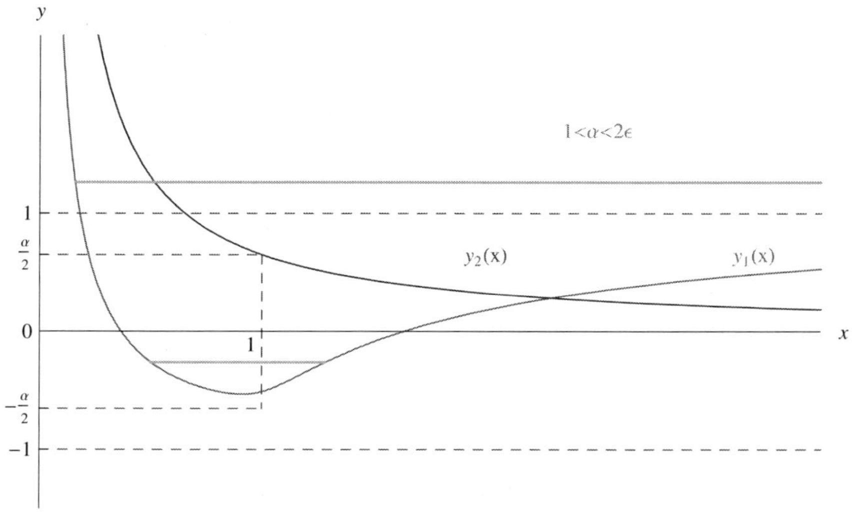

Figure 97. The effective potential for heavy shells at $\alpha/2 < \varepsilon = 1 < \alpha$ and $\sigma_{\text{in}} = +1$.

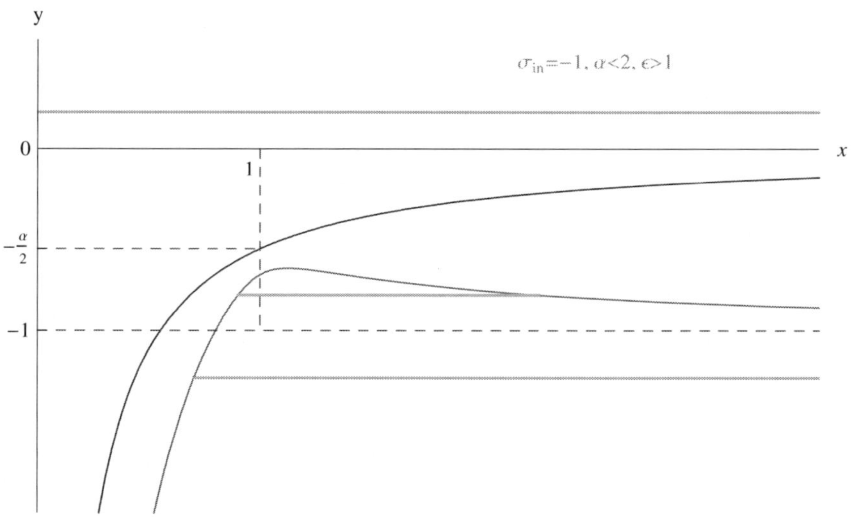

Figure 98. The effective potential for light shells at $\alpha < 2\varepsilon$ at $\sigma_{\text{in}} = -1$.

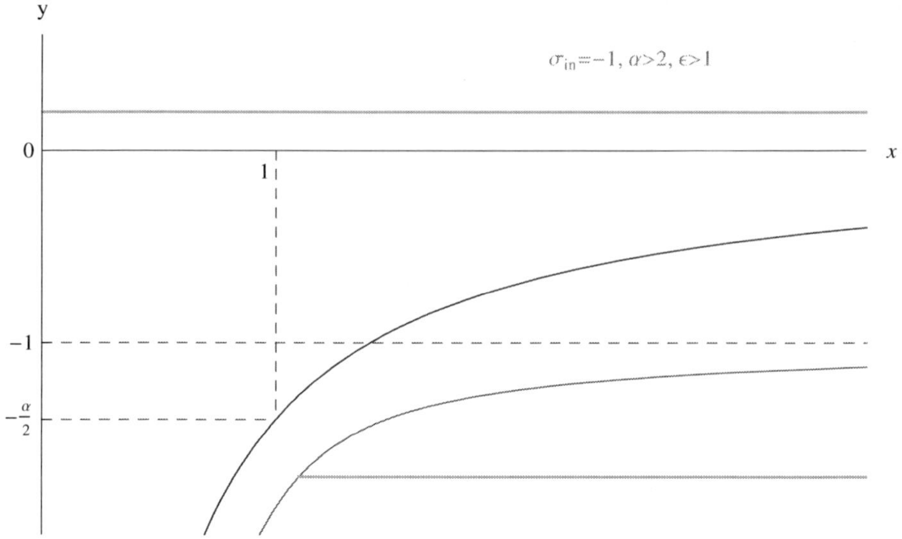

Figure 99. The effective potential for heavy shells at $\alpha > 2\varepsilon$ at $\sigma_{\rm in} = -1$.

lie below the potential curve. There can be either the bound motion with two turning points (for light shells), or the unbound motion with one turning point (both for light and heavy shells). The Carter–Penrose diagrams are shown in Fig. 100.

10.3. The Case $\varepsilon < 1$

At last, let us consider the case $\varepsilon < 1$, when the inner part of the combined manifold represents the metrics of the Reissner–Nordström black hole. Its structure is much more sophisticated than what we dealt with before. The mail feature is that the Reissner–Nordström black hole space-time contains all possible types of regions, $R_\pm$ and $T_\pm$ ones. This is the reason why the pictures for the effective potential look so differently and unusually. Fortunately, not only the authors but the readers as well, all of us have enough experience (due to reading the preceding Sections) in order to omit the details and to construct easily the Carter–Penrose diagram (= global geometry) just looking at the corresponding picture for the effective potential.

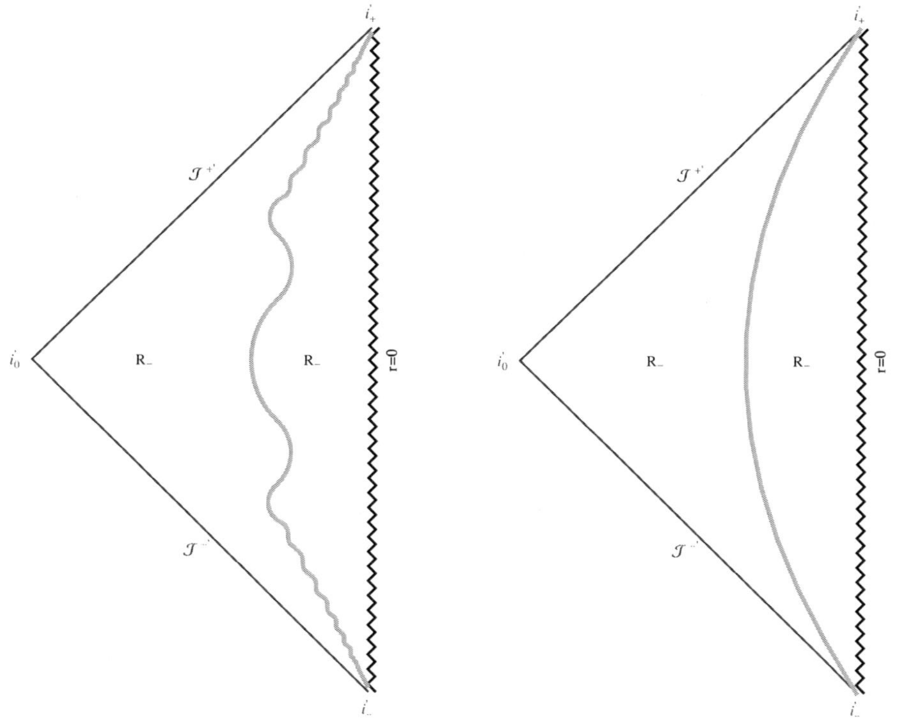

Figure 100. The Carter–Penrose diagram at $\varepsilon > 1$, $\sigma_{in} = -1$, $\alpha < 2$ for $-1 < \mu < 0$ (left panel) and, respectively, for $\alpha \gtrless 2$, $\mu < -1$ (right panel).

Consider, first, the case $\varepsilon^2 < \alpha^2/4 < \alpha^2 < 1$. Below in Fig. 101 is shown the effective potential $y = y_1$ together with the curves y_2 and $\tilde{y}_2 = -y_2$.

We enumerated the possible types of the trajectories as $\mu_1 < \mu_2 < \mu_3 < \mu_4 < \mu_5 < \mu_6$. Since we confined ourselves to only positive values of the total mass m_{out}, there should be $1 + \mu\alpha > 0$, so $\mu > \mu_0 = 1/\alpha$. As the first step, we consider the unbound motion, $\mu < -1$, with no turning points, thus, the range for μ_1 is $-1/\alpha < \mu_1 < -1$. For negative values of μ the outer part of the combined manifold could be, surely, the Reissner–Nordström black holes, $-(1-\varepsilon^2)/\alpha < \mu_1$, as well as the extreme black hole, $\mu_1 = -(1-\varepsilon^2)/\alpha$, or the space-time with naked singularity, $\mu_1 < -(1-\varepsilon^2)/\alpha$. The Carter–Penrose diagrams are shown below in Figs. 102 and 103.

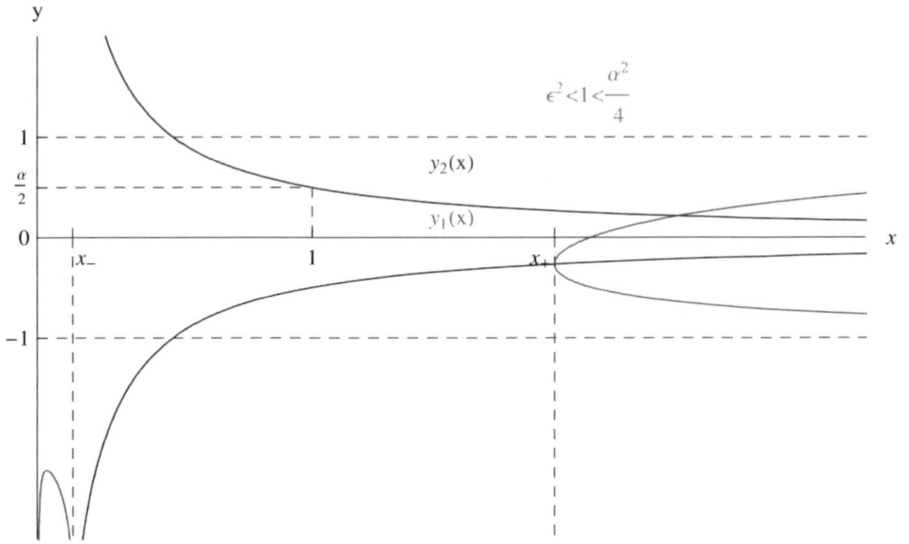

Figure 101. The effective potential for the case $\varepsilon^2 < \alpha^2/4 < \alpha^2 < 1$.

Next, the interval $-1 < \mu = \mu_2 < -\alpha/(2x_+)$, where we have the bound motion that starts from the singularity at zero radius in the R_+–region ($\sigma_{\text{in}} = +1$) of the inner metrics, and ends at the turning point in the R_-–region ($\sigma_{\text{in}} = -1$) outside the event horizon (x_+) of the inner metrics. What concerns the outer metrics, $\sigma_{\text{out}} = -1$ everywhere there. The situation is the same as for μ_1: we may have all three representatives of the Reissner–Nordström space-time. The conformal diagrams are the following, Figs. 104 and 105.

In the interval $-\alpha/(2x_+) < \mu = \mu_3 < 0$, everywhere on the trajectories, $\sigma_{\text{in}} = +1$ in the inner part and $\sigma_{\text{out}} = -1$ in the outer part of the combined manifold. The motion is bound with one turning point. Below in Figs. 106 and 107 are shown the Carter–Penrose diagrams for different types of the outer metrics.

Let us come to the positive values of μ. Now $\sigma_{\text{in}} = +1$ everywhere on the trajectories, and the metrics outside is that of the Reissner–Nordström black hole. There are three different intervals for μ. If $0 < \mu = \mu_4 < \alpha/(2x_2)$ (where x_2 is the intersection point of the right hand branch of the effective potential and the curve of changing σ_{out}, $y = y_2$ $(x_2 > x_+)$), then $\sigma_{\text{out}} = -1$ everywhere on the trajectories. The Carter–Penrose diagram in this case differs from the preceding one (for the black hole outside) only by the interchange in the positions of the

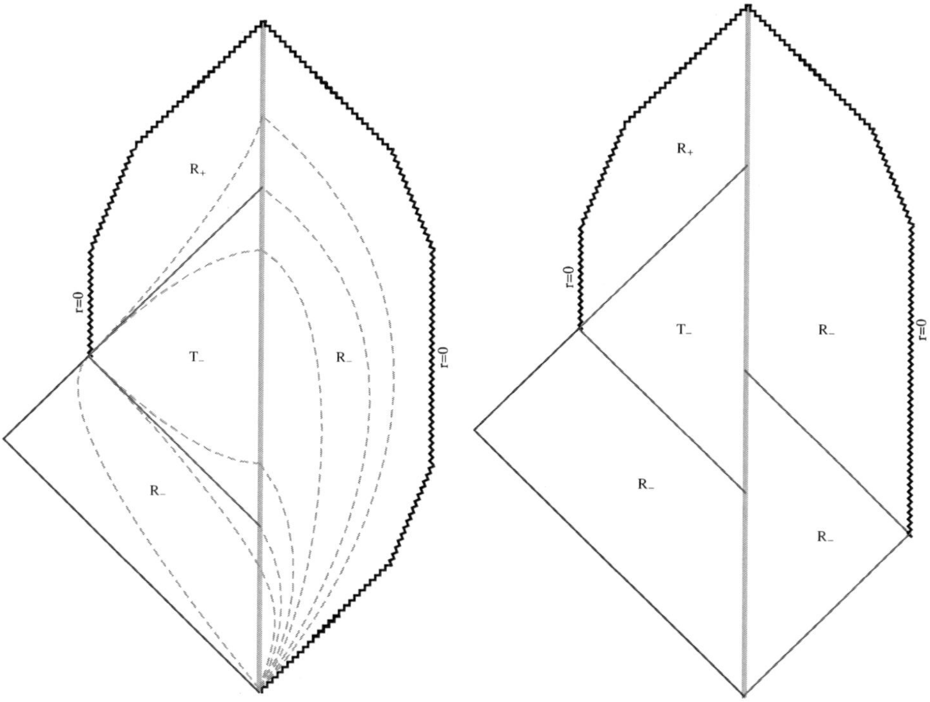

Figure 102. The Carter–Penrose diagram at $\varepsilon^2 < \alpha^2/4 < \alpha^2 < 1$ for $-1/\alpha < \mu < -(1-\varepsilon^2)/\alpha < -1$ (left panel) and, respectively, for $-1/\alpha < \mu = -(1-\varepsilon^2)/\alpha < \mu < -1$ (right panel).

horizons (due to changing sign of μ). The motion is bound with one turning point. If $\alpha/(2x_2) < \mu = \mu_5 < 1$, then the turning point moves to the R_+–region outside the event horizon of the outer metrics. Finally, for $\mu = \mu_6 > 1$ we have the unbound motion, the shell collapses from the past temporal infinity in the R_+–region of the outer metrics into the R_-–region near singularity. Surely, there exists also the time-reversal motion. The conformal diagrams for these three intervals of μ are shown in Figs. 108 and 109, respectively.

The next step is changing the relations between ε and α to the range $\alpha^2/4 < \varepsilon^2 < \alpha^2 < 1$. The effective potential becomes drastically different, as shown in Fig. 110.

What is essential here? There appears the turning point near singularity.

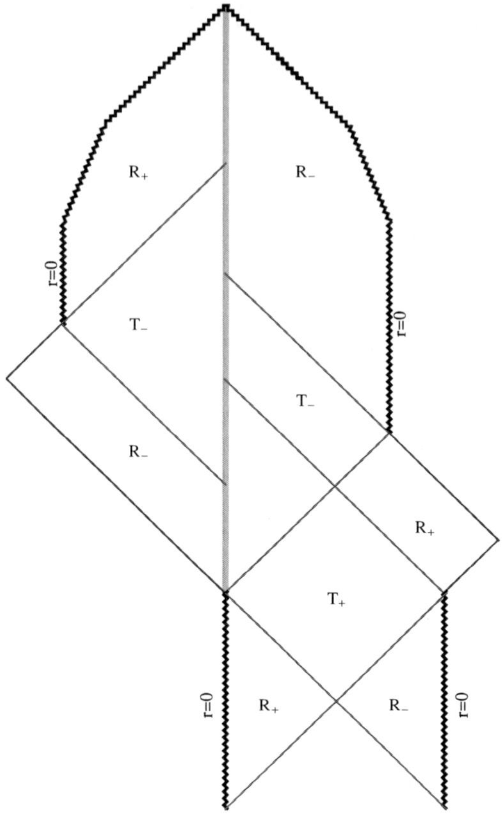

Figure 103. The Carter–Penrose diagram at $\varepsilon^2 < \alpha^2/4 < \alpha^2 < 1$ for $-1/\alpha < -(1-\varepsilon^2)/\alpha < -1 < \mu < -1$.

Consequently, all the Carter–Penrose diagrams should be replaced. One more thing: the horizontal line $\mu = \mu_0 = -1/\alpha$, i. e., $m_{\text{out}} = 0$, may lie lower as well as above the intersection point of the left hand branch of the effective potential with the curve $\tilde{y}_2$, i. e., $y_1 = \tilde{y}_2 = -y_2 = -\alpha/(2x_-)$. So, let $\mu = \mu_1 < -\alpha/(2x_-)$. Below in Figs. 111 and 112 are the conformal diagrams for all three types of the outer metrics. Remember, that everywhere on the trajectories $\sigma_{\text{in}} = -1$ and $\sigma_{\text{out}} = -1$.

For higher values of μ, namely, $-\alpha/(2x_-) < \mu = \mu_2 < -1$, we have slightly different diagrams inside the shell: now the turning point lies in the R_+–region

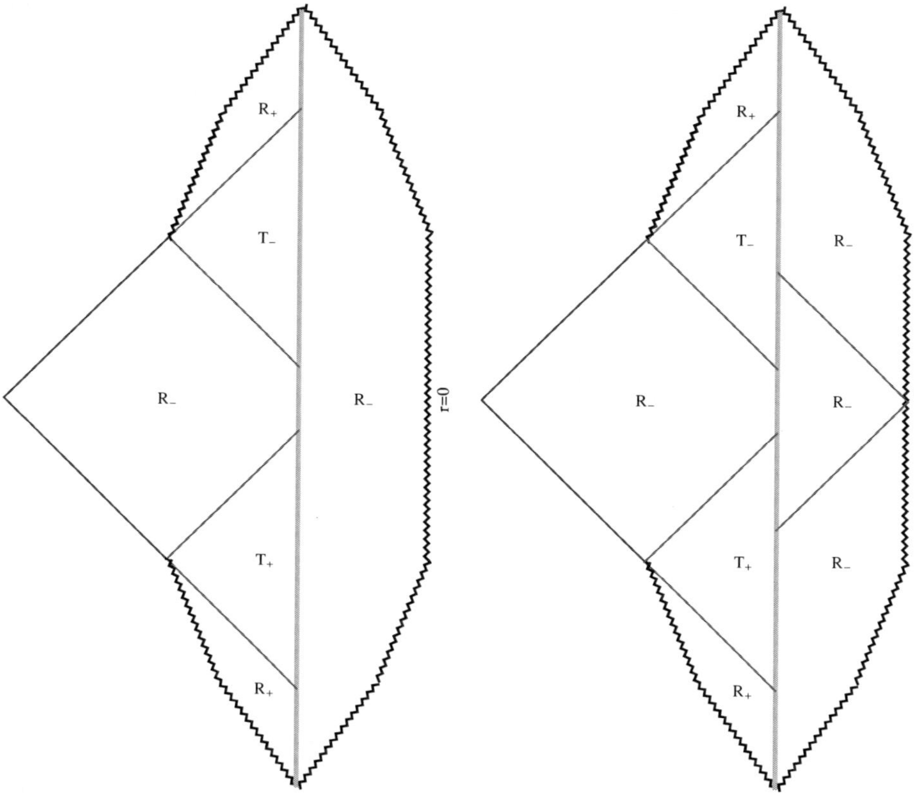

Figure 104. The Carter–Penrose diagram at $\varepsilon^2 < \alpha^2/4 < \alpha^2 < 1$ for $-1 < \mu < -(1-\varepsilon^2)/\alpha < \alpha(2x_+)$ (left panel) and, respectively, for $-1 < \mu + -(1-\varepsilon^2)/\alpha < \alpha/(2x_+)$ (right panel).

near singularity, but outside the shell everything remains the same, as shown in Figs. 113 and 114.

After shifting μ further up, into the region $-1 < \mu = \mu_3 < -\frac{\alpha}{2x_+}$, we get the bound motion with two turning points and with opposite signs of σ_{in} at the ends, but the same $\sigma_{\text{out}} = -1$. The Carter–Penrose diagrams are the following in Figs. 115 and 116.

For even higher, but still negative, values of μ, namely, $-\alpha?(2x_+) < \mu = \mu_4 < 0$, we have the motion of the same type as that considered above, but now

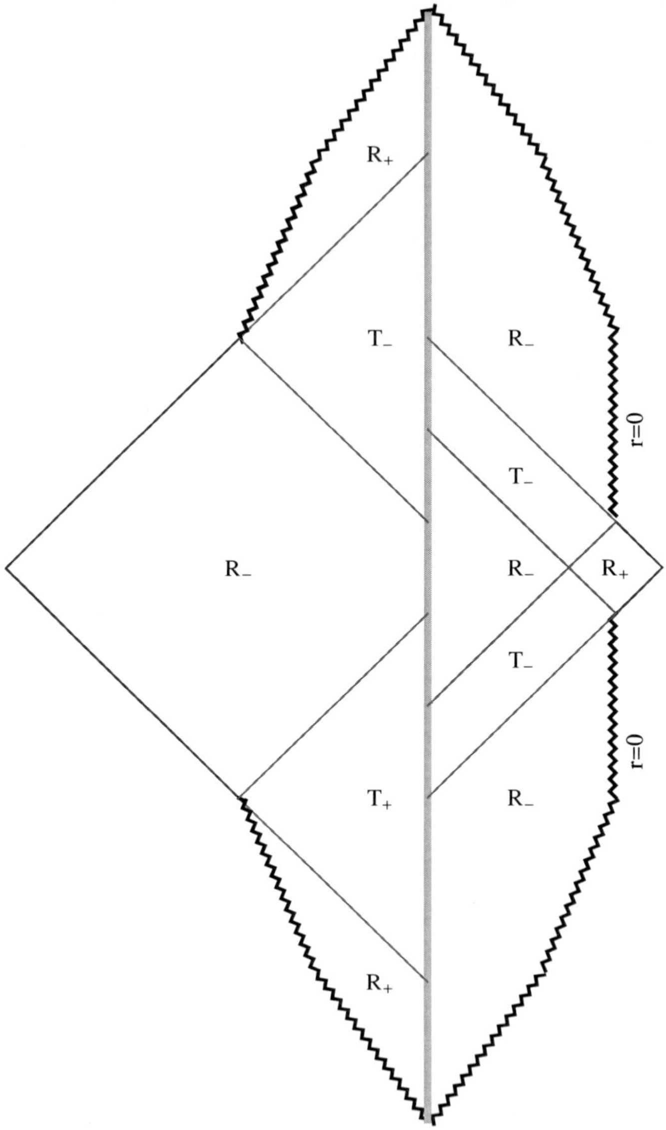

Figure 105. The Carter–Penrose diagram at $\varepsilon^2 < \alpha^2/4 < \alpha^2 < 1$ for $-1 < -(1-\varepsilon^2)/\alpha < \alpha/(2x_+)$.

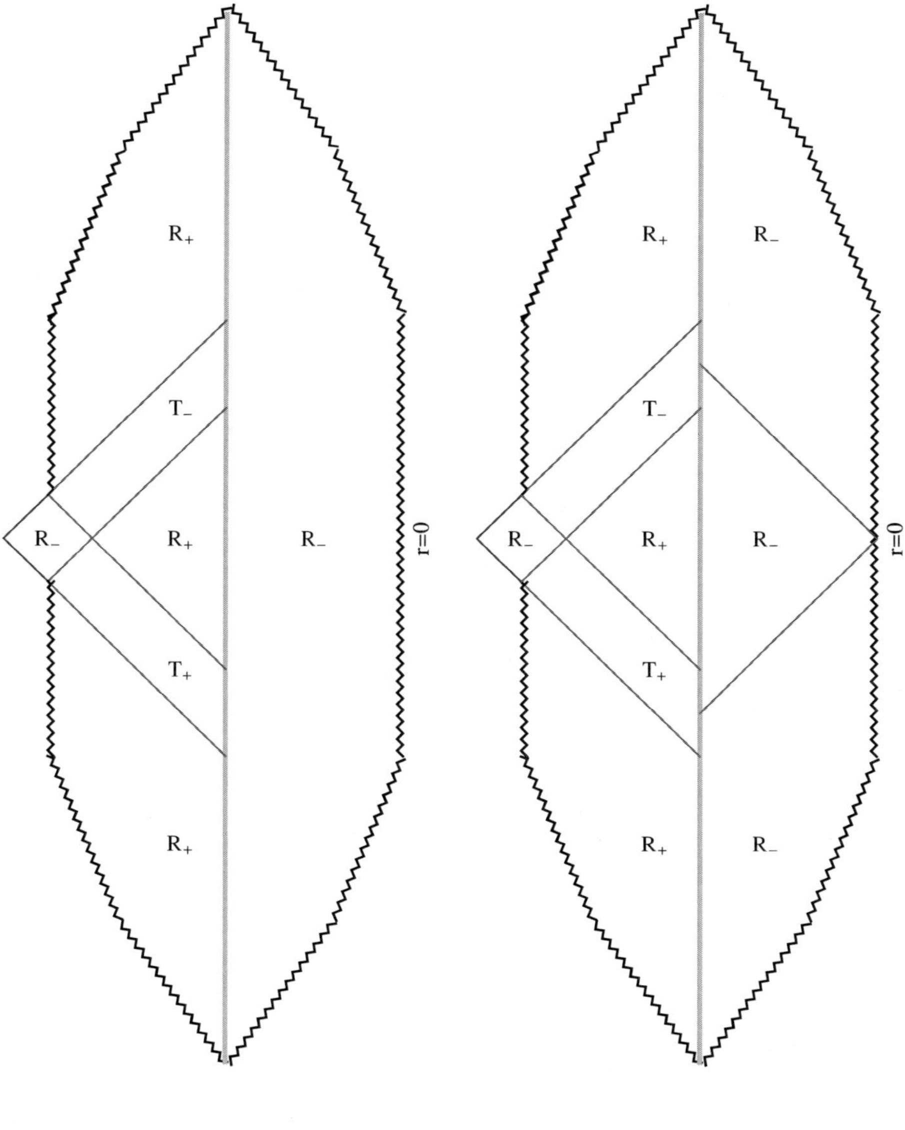

Figure 106. The Carter–Penrose diagram at $\varepsilon^2 < \alpha^2/4 < \alpha^2 < 1$ for $-1 < -\alpha/(2x_+) < \mu < 0$ (left panel) and, respectively, for $-1 < -\alpha/(2x_+) < \mu < 0$ (right panel).

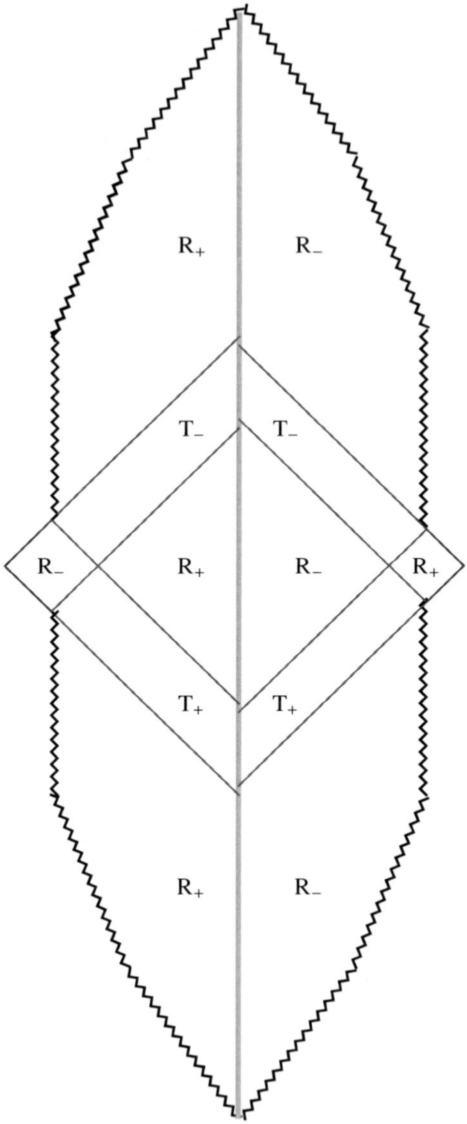

Figure 107. The Carter–Penrose diagram at $\varepsilon^2 < \alpha^2/4 < \alpha^2 < 1$ for $-1 < -\alpha/(2x_+) < \mu < 0$.

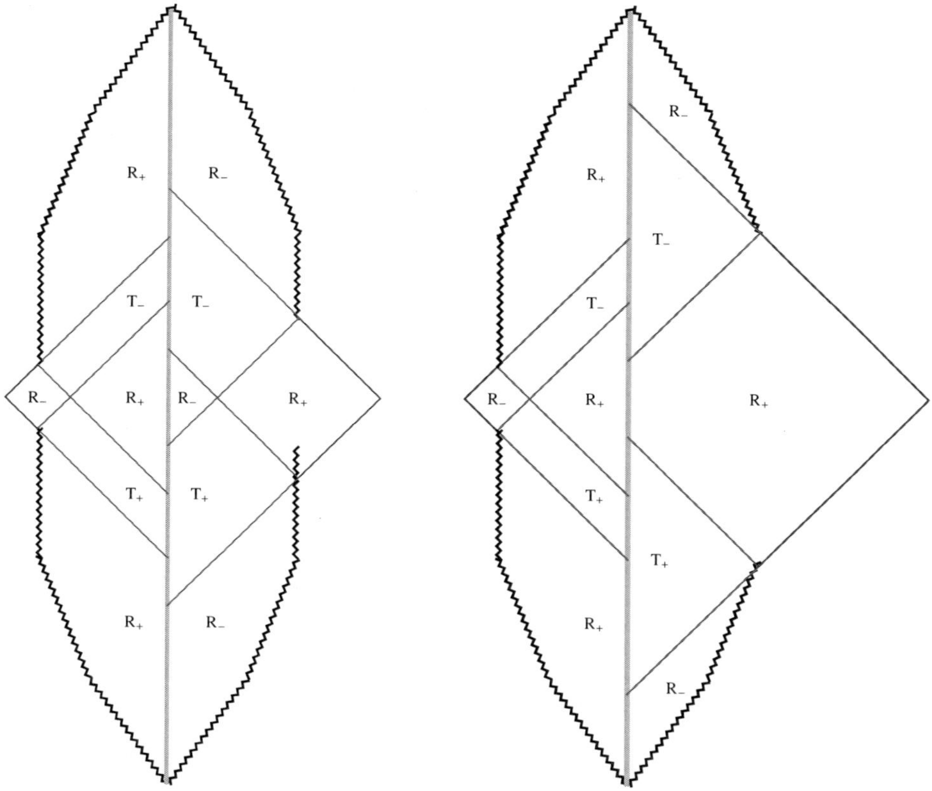

Figure 108. The Carter–Penrose diagram at $\varepsilon^2 < \alpha^2/4 < \alpha^2 < 1$ for $0 < \mu < \alpha/(2x_2)$ (left panel) and, respectively, for $\alpha/(2x_2) < \mu < 1$ (right panel).

both $\sigma' s$ do not change their signs, $\sigma_{\rm in} = +1$, $\sigma_{\rm out} = -1$. See the corresponding diagrams in Figs. 117 and 118.

Let now $\mu > 0$. Outside the shell there cam be only the Reissner–Nordström metric. One should consider three different intervals:

(a) $0 < \mu = \mu_5 < \alpha/(2x_2)$ - bound motion with two turning points and $\sigma_{\rm out} = -1$ everywhere;

(b) $\alpha/(2x_2) < \mu = \mu_6 < 1$ - bound motion with two turning points and opposite values of $\sigma_{\rm out}$ at the ends ($\sigma_{\rm out} = -1$ near the singularity and $\sigma_{\rm out} = +1$ outside the event horizon);

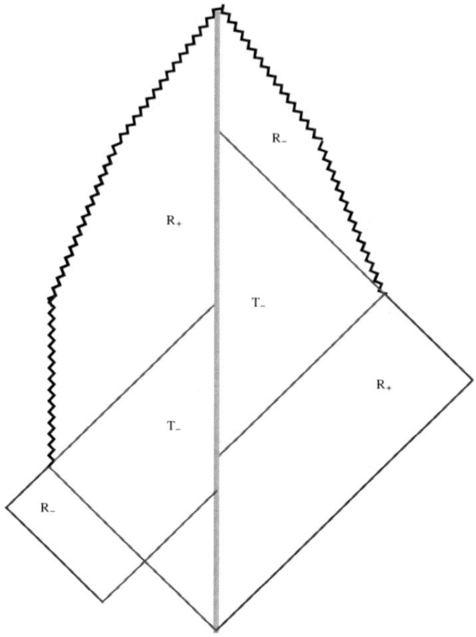

Figure 109. The Carter–Penrose diagram at $\varepsilon^2 < \alpha^2/4 < \alpha^2 < 1$ for $\mu > 1$.

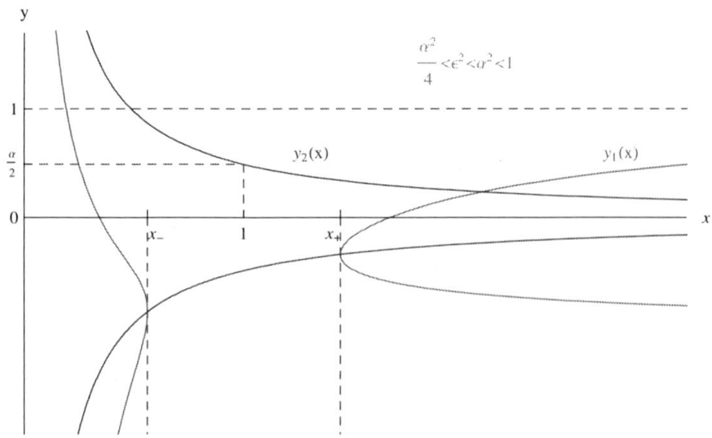

Figure 110. The effective potential for the case $\alpha^2/4 < \varepsilon^2 < \alpha^2 < 1$.

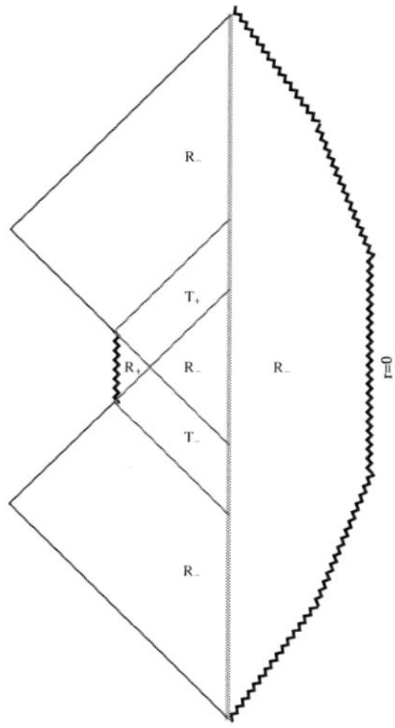

Figure 111. The Carter–Penrose diagram at $\alpha^2/4 < \varepsilon^2 < \alpha^2 < 1$ for $\mu > 1$.

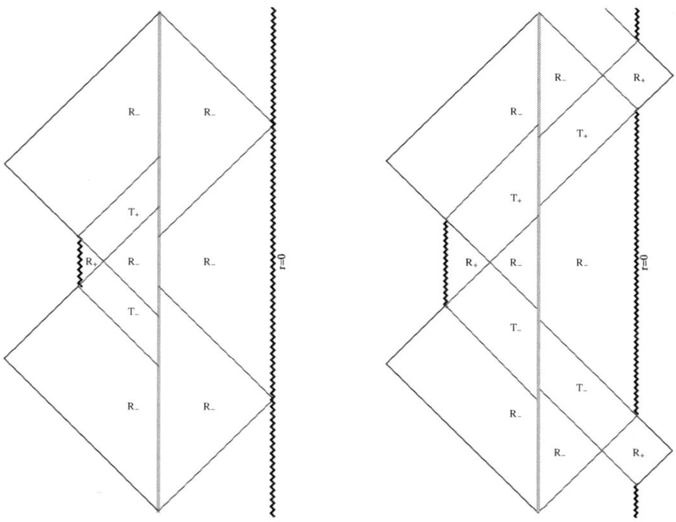

Figure 112. The Carter–Penrose diagram at $\alpha^2/4 < \varepsilon^2 < \alpha^2 < 1$ for $0 < \mu < \alpha/(2x_2)$ (left panel) and, respectively, for $\alpha/(2x_2) < \mu < 1$ (right panel).

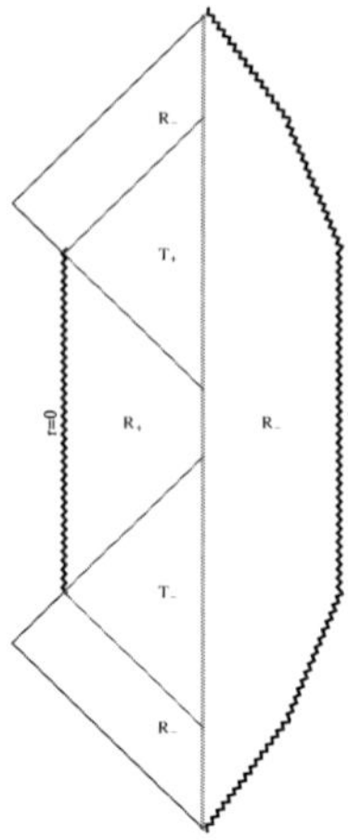

Figure 113. The Carter–Penrose diagram $\alpha^2/4 < \varepsilon^2 < \alpha^2 < 1$ for $-\alpha/(2x_-) < \mu = \mu_2 < -1$.

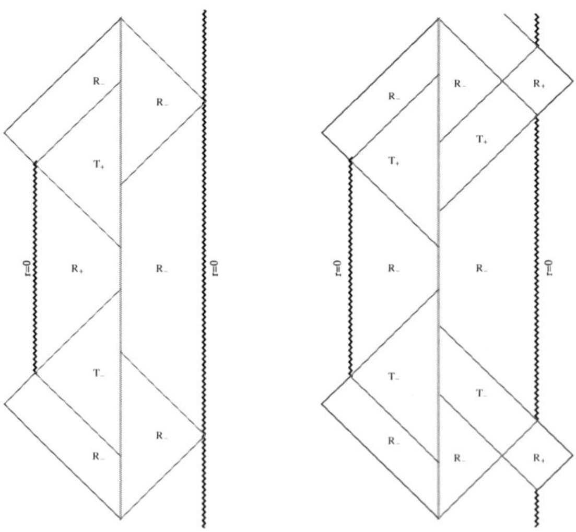

Figure 114. The Carter–Penrose diagram at $\alpha^2/4 < \varepsilon^2 < \alpha^2 < 1$ for, respectively, $-\alpha/(2x_-) < \mu < \mu_2 < -1$ (left panel) and for $-\alpha/(2x_-) < \mu_2 < \mu < -1$ (right panel).

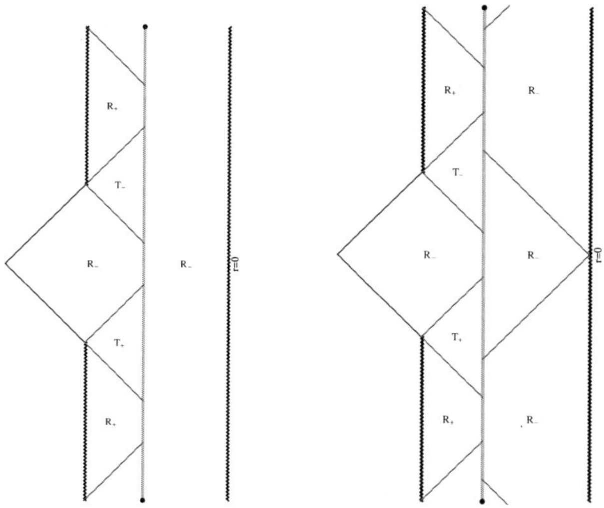

Figure 115. The Carter–Penrose diagram at $\alpha^2/4 < \varepsilon^2 < \alpha^2 < 1$ for, respectively, $-1\mu < -(1-\varepsilon)/\alpha) < -\alpha/(2x_-)$ (left panel) and for $-1\mu = -(1-\varepsilon)/\alpha) < -\alpha/(2x_-)$ (right panel).

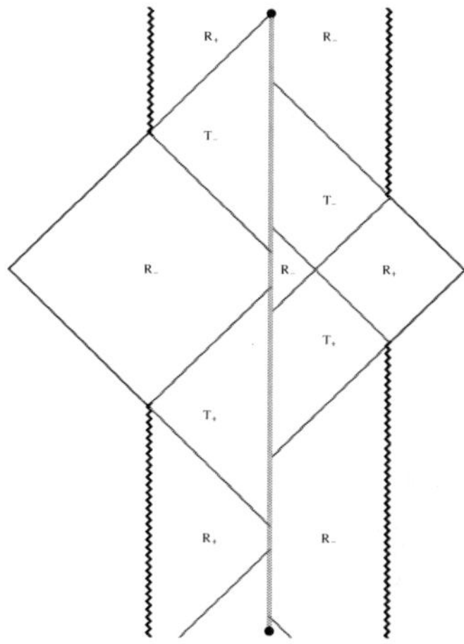

Figure 116. The Carter–Penrose diagram at $\alpha^2/4 < \varepsilon^2 < \alpha^2 < 1$ for $-1 < -(1-\varepsilon)/\alpha) < \mu < -\alpha/(2x_-)$.

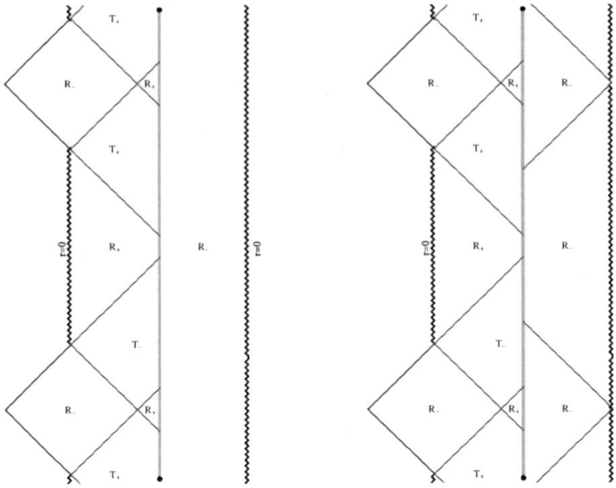

Figure 117. The Carter–Penrose diagram at $\alpha^2/4 < \varepsilon^2 < \alpha^2 < 1$ for, respectively, $-\alpha/(2x_-) < \mu < -(1-\varepsilon)/\alpha) < 0$ (left panel) and for $-\alpha/(2x_-) < \mu = -(1-\varepsilon)/\alpha) < 0$ (right panel).

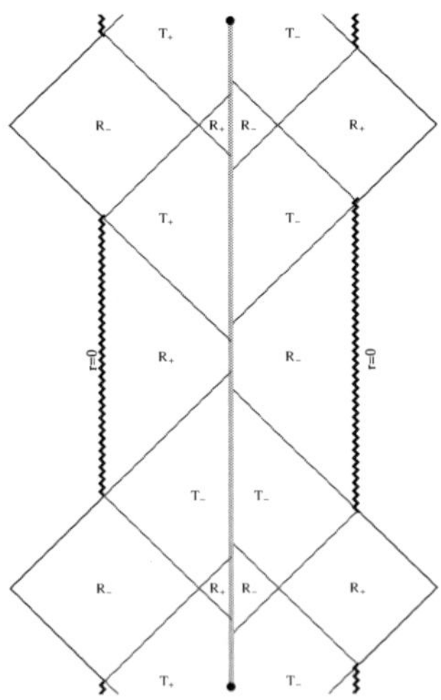

Figure 118. The Carter–Penrose diagram at $\alpha^2/4 < \varepsilon^2 < \alpha^2 < 1$ for $-\alpha/(2x_-) < -(1-\varepsilon)/\alpha) < \mu < 0$.

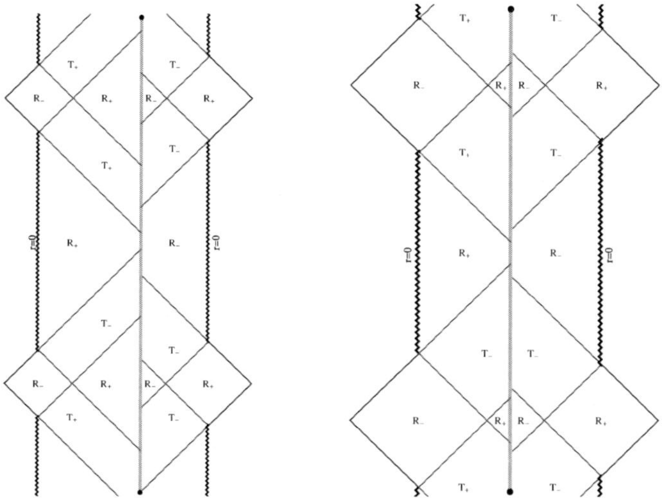

Figure 119. The Carter–Penrose diagram at $\alpha^2/4 < \varepsilon^2 < \alpha^2 < 1$ for, respectively, $0 < \mu < \alpha/(2x_+) < 1$ (left panel) and for $\alpha/(2x_+) < \mu < 1$ (right panel).

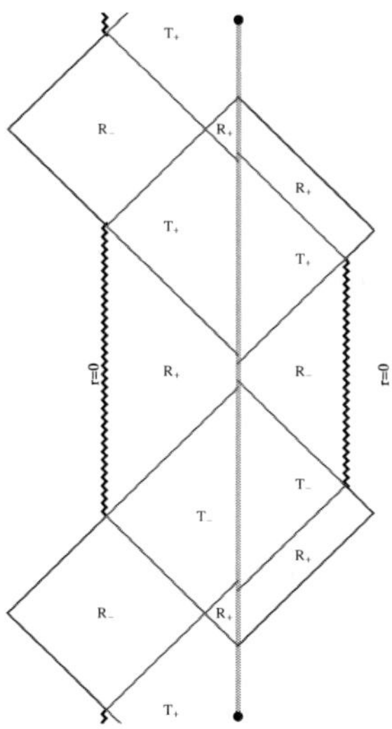

Figure 120. The Carter–Penrose diagram at $\alpha^2/4 < \varepsilon^2 < \alpha^2 < 1$ for $\mu > 1$.

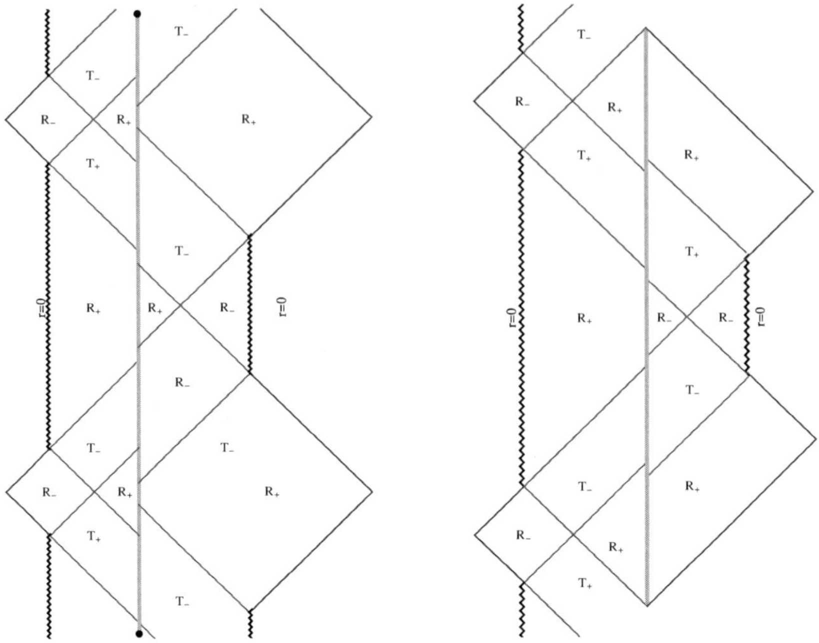

Figure 121. The Carter–Penrose diagram at $\alpha^2/4 < \alpha^2 < \varepsilon^2 < 1$ for, respectively, $\alpha/(2x_1) < \mu < 1$ (left panel) and for $1 < \alpha/(2x_1) < \mu$ (right panel).

(c) $\mu = \mu_7 > 1$ - unbound motion with one turning point and changing sign of σ_{out} ($\sigma_{\text{out}} = -1$ at the turning point). Remember, that $\sigma_{\text{in}} = +1$ everywhere.

See the corresponding diagrams in Figs. 119 and 120.

Let us now interchange ε and α: $\alpha^2/4 < \alpha^2 < \varepsilon^2 < 1$. There appears the intersection point, x_1, of the left hand branch of the effective potential with the curve y_2 and, therefore, the possibility for the shell to move without changing $\sigma_{\text{out}} = +1$. In what follows we will show only those new conformal diagrams when $\mu > \alpha/(2x_1)$. Since the intersection point $y_1 = y_2$ may lie below as well as above the line $y = 1$, we get two different global geometries, shown in Fig. 121.

Further increasing the values of α does not give us the new types of global geometries. The reader can easily find them among the already drawn diagrams.

References

[1] W. Israel, *Phys. Rev*, 153, 1388 (1967).

[2] L. D. Landau and E. M. Lifshitz, *The classical theory of fields* (Oxford: Pergamon Press, 1975).

[3] V. A. Berezin, V. A.Kuzmin, I. I. Tkachev, *Phys. Rev. D* 36, 2919 (1987).

[4] V. A. Berezin, V. A.Kuzmin, I. I. Tkachev, *Phys. Lett. B* 120, 91 (1983).

[5] V. A. Berezin, V. A.Kuzmin, I. I. Tkachev, *Phys. Lett. B* 124, 479 (1983).

[6] V. A. Berezin, V. A.Kuzmin, I. I. Tkachev, *Phys. Lett. B* 130, 23 (1983).

[7] V. A. Berezin, V. A.Kuzmin, I. I. Tkachev, *Pis'ma ZhETF* 41, 446 (1985).

[8] V. A. Berezin, V. A.Kuzmin, I. I. Tkachev, *ZhETF*, 86, 785 (1984).

[9] V. A. Berezin, V. A.Kuzmin, I. I. Tkachev, *ZhETF* 93, 1159 (1987).

[10] V. Berezin, V. Dokuchaev, Yu. Eroshenko, A. Smirnov, *Class. Quant. Grav.* 22, 4443 (2005).

[11] V. A. Berezin and V. I. Dokuchaev, *Global space-time geometry with the charged shell.* Preprint INR 1343/2013 (2013).

Authors' Contact Information

Dr. Victor Aleksandrovich Berezin,
Institute for Nuclear Research of the Russian Academy of Sciences
60th October Anniversary Prospect 7a, 117312 Moscow, Russia
Tel: +7-903-561-6131
Email: berezin@inr.ac.ru

Dr. Vyacheslav Ivanovich Dokuchaev,
Institute for Nuclear Research of the Russian Academy of Sciences
60th October Anniversary Prospect 7a, 117312 Moscow, Russia
Tel: +7-906-045-8561
Email: dokuchaev@inr.ac.ru